Security and Political Discourse

Security and Political Discourse

How Does the Naming of Challenges as a Security Issue Determine Political Practice?

By

Paulina Polko

BRILL

LEIDEN | BOSTON

The project was co-financed by funds from the Polish state budget, awarded by the Minister of Science within the framework of the Excellent Science II Programme.

Cover illustration: "Knowledge." Image by (user) lolloj. 2015. Available via iStock.

The Library of Congress Cataloging-in-Publication Data is available online at https://catalog.loc.gov
LC record available at https://lccn.loc.gov/2025036409

Typeface for the Latin, Greek, and Cyrillic scripts: "Brill". See and download: brill.com/brill-typeface.

ISBN 978-90-04-74540-7 (hardback)
ISBN 978-90-04-74541-4 (e-book)
DOI 10.1163/9789004745414

This book is printed on acid-free paper and produced in a sustainable manner.

Contents

Figures and Tables

Figures

Tables

Abbreviations

APEC	*Asia-Pacific Economic Cooperation*
AWS	Akcja Wyborcza "Solidarność" [*"Solidarity" Electoral Action*]
BBWR	Bezpartyjny Blok Wspierania Reform [*Nonpartisan Reform Support Bloc*]
CBA	Centralne Biuro Antykorupcyjne [*Central Anticorruption Bureau*]
CBŚ/CBŚP	Centralne Biuro Śledcze (Policji) [*Central Police Investigation Bureau*]
CSCE	*Conference on Security and Cooperation in Europe*
CSS	*Critical Security Studies*
EC	*European Commission*
EEC	*European Economic Community*
EGD	*European Green Deal*
EU	*European Union*
FPD	Forum Prawicy Demokratycznej [*Forum of the Democratic Right*]
FR	Federacja Rosyjska [*Russian Federation*]
ILO	*International Labour Organisation*
KLD	Kongres Liberalno-Demokratyczny [*Liberal Democratic Congress*]
KO	Koalicja Obywatelska [*Civic Coalition*]
KO "S"	Komitet Obywatelski "Solidarność" [*"Solidarity" Citizens' Committee*]
KP	Koalicja Polska [*Polish Coalition*]
LiD	Lewica i Demokraci [*Left and Democrats*]
LPR	Liga Polskich Rodzin [*League of Polish Families*]
MON	Ministerstwo Obrony Narodowej [*Ministry of Defense*]
NATO	*North Atlantic Treaty Organization*
NSZZ	"Solidarność" – Niezależny Samorządny Związek Zawodowy "Solidarność" [*Independent Self-Governing Trade Union "Solidarity"*]
OSCE	*Organization for Security and Cooperation in Europe*
PC	Porozumienie Centrum [*Center Agreement*]
PChD	Partia Chrześcijańskich Demokratów [*Christian Democrats*]
PiS	Prawo i Sprawiedliwość [*Law and Order*]
PO	Platforma Obywatelska [*Civic Platform*]
PPPP	Polska Partia Przyjaciół Piwa [*Polish Party of Beer Friends*]
PPN	Polska Partia Narodowa [*Polish National Party*]
PRL	Polska Rzeczpospolita Ludowa [*Polish Peoples Republic*]
PSL	Polskie Stronnictwo Ludowe [*Polish People's Party*]
PSL-PL	Polskie Stronnictwo Ludowe – Porozumienie Ludowe [*Polish People's Party – People's Agreement*]

PZPR	Polska Zjednoczona Partia Robotnicza [*Polish United Workers' Party*]
ROAD	Ruch Obywatelski Akcja Demokratyczna [*Civic Movement Democratic Action*]
ROP	Ruch Odrodzenia Polski [*Polish Rebirth Movement*]
RP	Rzeczpospolita Polska [*Republic of Poland*]
RWPG	Rada Wzajemnej Pomocy Gospodarczej [*Council for Mutual Economic Assistance*]
SD	Stronnictwo Demokratyczne [*Democratic Party*]
SdRP	Socjaldemokracja RP [*Social Democracy of the Republic of Poland*]
SLCh	Stronnictwo Ludowo-Chrześcijańskie [*People's Christian Party*]
SLD	Sojusz Lewicy Demokratycznej [*Democratic Left Alliance*]
SP	Solidarna Polska [*Solidarity Poland*]
SP	Suwerenna Polska [*Sovereign Poland*]
SZ RP	Siły Zbrojne Rzeczypospolitej Polskiej [*Polish Armed Forces*]
UD	Unia Demokratyczna [*Democratic Union*]
UN	*United Nations*
UNHCR	*United Nations High Commissioner for Refugees*
UP	Unia Pracy [*Labour Union*]
UPR	Unia Polityki Realnej [*Union of Real Politics*]
UW	Unia Wolności [*Union of Freedom*]
WFP	*World Food Programme*
WHO	*World Health Organization*
WNP	Wspólnota Niepodległych Państw [*Union of Independent States*]
WOT	Wojska Obrony Terytorialnej [*Territorial Defence Forces*]
WP	*Warsaw Pact*
ZChN	Zjednoczenie Chrześcijańsko-Narodowe [*Christian National Union*]
ZL	Zjednoczona Lewica [*United Left*]
ZP	Zjednoczona Prawica [*United Right*]
ZSSR	Związek Socjalistycznych Republik Radzieckich [*Soviet Union*]

Introduction: Security as a Category Shaping Political Discourse

Although security has consistently been recognized as one of the fundamental needs and aspirations of humans, the 21st century and the threats it brings have made this concept a category of particular paramount importance. The ongoing securitization of many areas of human activity, previously unanalyzed from the security perspective, has significantly increased the role of this concept in political discourse, and its *alter ego* – threat, or rather the fear of its occurrence – has become an important rhetorical device and persuasive meta-operator (Awdiejew, 2004: 64–79; Awdiejew, 2007: 142–152) building this discourse.

Security now shapes state policy. Society strictly evaluates the effects of this policy in light of its expectations. The gap between these expectations and the state's capacity can be significant. The government primarily views its role as mitigating threats to the state as a whole and secondarily, to a lesser extent, in addressing challenges and creating opportunities for the individual development of human beings. This classic approach, typical of the post-war bipolar world of the latter half of the 20th century, fails to reflect the expansion of the security category beyond the strictly military sphere, which is characteristic of the 21st century. "Here, the focus of security issues shifts to individuals and their social groups, who should be guaranteed freedom, sovereignty, prosperity, and development" (Urbanek, 2013: 11). In a world where threats are losing their dominant political-military character in favor of social, economic, demographic, environmental, and health-related threats, the list of citizens' expectations towards the state does not seem excessive.

The economy and demography, along with the natural environment and its threats, social security, health and medical progress, the development of modern technologies, gender issues and equal rights, and education, are all evaluated through the prism of security. Security procedures and experts in their implementation prevail. I. Fischhendler and D. Katz (2013: 324) discuss the use of "security jargon" in dialogues about sustainable development while analyzing issues such as climate and its changes, ecosystems and their functioning, energy, and the availability of food and drinking water. They highlight that the term security is linked to concepts and phenomena considered socially significant, which may be at risk. Participants in public discourse, including politicians, employ this to elevate issues that have not always received adequate attention. There is a prevailing belief that connecting climate challenges with national and international security grants them a "significant position both in the discourse itself and in the subsequent decision-making process" (Hale, 2010: 257).

Speech acts have constitutive meaning. The Copenhagen School of Security indicates that message recipients consider a phenomenon to be a threat to security when it is identified as such by the audience (Buzan et al., 1998). According to B. Buzan, the scare was successful and contributed to the creation of concepts such as environmental security (Graeger, 1996; Allenby, 2000; Trombetta, 2008), water security (Tarlock and Wouters, 2009), food security (World Bank, 1986), and energy security (Kruyt et al., 2009). The Riskification Theory, developed on this basis in 2012 by O. Corry, suggests that security issues are used to provide decision-makers with tools not only to combat threats in emergency situations but also to manage potential risks in quite ordinary times (Corry, 2012).

Securitization (and riskification) of increasingly new areas leads to violations of the rules of the normal political decision-making process, placing topics and problems related to security outside "normal politics" and even public debate. It removes them from being submitted to society's judgment as issues that "cannot be voted on" because they are too important for the public and serve to ensure its well-being (Schmitt, 1985; Williams, 2003) and gives them priority not only in terms of their position in discourse, but also in the sphere of decisions and actions (Balzacq, 2005; Wæver, 1995; Bethurem, 2002). To what extent does the securitization of various areas of life increase their importance and facilitate taking necessary actions (e.g., in the field of environmental protection), and to what extent is it just a rhetorical device and a marketing trick of little significance (Levy, 1995; Deudney, 1990a)? Isn't linking every aspect of human life and activity with security issues a double-edged sword that drains this concept of any meaning (Deudney, 1990ab; Stern and Ojendal, 2010; Ciută, 2010)? Can security make it possible to extend the use of extraordinary measures to contain low-risk threats indefinitely? (Corry, 2012) When should desecuritization take place? And if this is not happening, perhaps – following Rita Floyd's advice – we should examine the intentions of those calling for the securitization of subsequent areas (whether security is not just a populist tool for them) and carefully assess the threats they indicate in terms of their existential dimension? These questions concern contemporary researchers of both security as a dynamically developing phenomenon, as well as analysts of discourse and the rhetorical tools used in it.

This superior position of security over other key categories has several serious consequences, including misunderstandings and changes in the meaning of traditional concepts. It compels us to redefine the phenomena and processes described earlier and adopt a new perspective on their assessment, even if supporters of universal securitization themselves have doubts about the validity of such an approach. Ultimately, it alters politics itself and the associated discourse, which is a significant component of it.

In recent years, security has become a persuasive meta-operator that is meant to strengthen and reorganize the message. Quantitative research shows that while security issues have occupied a similar part of prime ministers' election programs and exposés over the last 30 years, the use of the term security itself has increased several dozen times, and the word frequently appears in contexts far removed from security.

Security extends beyond its lexical meaning and takes on roles beyond strictly informational ones. Often responsible for the persuasive and emotive aspects of the message, it justifies difficult decisions and explains a state of higher necessity or emergency (we must do something or, on the contrary, we cannot do something due to security reasons). It is also employed to instill fear in the recipients of the message, and once they reach the "desired" level of fear, it aims to build relationships (understanding) with them and, consequently, create a favorable image of the securitizing actor (as one who identifies the threat and intends to counteract it).

Thus, the word "security" plays an important conative (impressive) role in contemporary political and public discourse. However, it loses its basic meaning and connection to the issues that are vital for addressing security challenges. A classic example of such a situation may be the potential relocation of refugees from the Middle East and Africa to Poland and its presentation in the 2015 campaign, as well as the discourse on ways to solve the crisis on the eastern border of the Schengen area in 2021. Political decisions not to accept refugees were made not based on a rational assessment of the situation, including the real threat associated with accepting a small group of people, but rather on an emotional, fearful, and alarming public discussion. Fear – the *alter ego* of security – is today an important element of political discourse, with no indication that this situation will change. This creates a need to treat the category of security differently, considering its discursive meaning. The influence of security discourse on political decisions, including the construction of the security system, is expected to increase rather than decrease in the coming years, with subjective assessments of the state of security (threats) serving as the basis for significant decisions related to ensuring security (elimination, minimization, or counteraction of threats). Therefore, it seems justified to analyze political discourse in the realm of security and assess its potential impact on systemic solutions implemented by politicians.

The aim of this book is to analyze political discourse on security in Poland after 1989, highlighting securitization processes, their extent, and intensity across various security sectors. Poland, one of the largest countries in Central and Eastern Europe, emerged as a leader in the political transformations of the 1980s, as well as in the accession processes to NATO and the European Union at the turn of the 20th and 21st centuries. It shaped its security policy in a way that mirrors the experiences of neighboring countries, facing similar threats and challenges as other post-socialist nations, such as Russian imperialism, the necessity to compensate for years of economic underdevelopment, negative population growth, a lack of energy resources, and the new challenge of international migration. The political discourse on security in Poland post-1989 reflects typical patterns observed in the region and reveals which aspects of security are salient for this part of Europe. Analyzing this discourse allows us to identify trends also present in other Central and Eastern European countries.

This book applies the theoretical frameworks of three theories that have emerged in the field of Critical Security Studies: Securitization Theory (ST), Just Securitization Theory (JST), and Riskification Theory (RT) to analyze securitization processes identified in selected areas of Polish political discourse after 1989.

The main problem raised during the work on this book was the question of the importance of security issues and the dynamics of securitization processes in Polish political discourse after 1989. Specific research problems related to the following:

1. Changes in the understanding of the concept of security and its threats over the last 30 years, reflected in political discourse;
2. Changes in the position of security issues in political discourse;
3. The level of securitization and riskification of the Polish political discourse;
4. Understanding and the manner of the use of security issues by different political parties and communities;
5. Fear and threatening as elements of political discourse;
6. Distinguishing between securitizations (extending the security category according to the Copenhagen School's assumptions), naming security issues without a securitization process, and tampering with the concept of security;
7. Desecuritization of selected areas previously recognized as exceptional issues due to belonging to the sphere of security.

The amount of material collected from the research necessitated narrowing down further analysis to selected elements of Polish political discourse. Therefore, the official programs of political parties participating in the elections were analyzed:

- parliamentary, from the years: 1989, 1991, 1993, 1997, 2001, 2005, 2007, 2011, 2015, 2019, 2023;
- presidential elections, from the years: 1990, 1995, 2000, 2005, 2010, 2015, 2020;

and:

- exposé of Polish prime ministers from 1989 (from Tadeusz Mazowiecki) to 2023 (Donald Tusk's third speech), speeches of the presidents of the Republic of Poland before the National Assembly and when assuming authority over the armed forces (from Wojciech Jaruzelski to Andrzej Duda);
- selected parliamentary debates on essential security issues (e.g. deploying Polish Armed Forces to Iraq, withdrawing troops from Afghanistan, accepting refugees' relocation, fighting the COVID-19 pandemic, supporting Ukraine and Ukrainian refugees in the fight against Russian aggression).

Speeches by individual politicians or press or television interviews were not analyzed. The assumption was based on the belief that it is crucial to examine what individual groups and the candidates they put forward have to offer in the area of security and how important these issues are compared to other issues raised during the campaign.

The book is divided into six chapters. The first chapter contains the theoretical basis of research, which refers to the evolution of understanding and defining security, particularly the changes initiated in this area by constructivists and the Copenhagen School, significantly expanding the category of security to new areas. It also presents critical approaches to the Securitization Theory, including Riskification and Just Securitization Theory, as well as the characteristics of political discourse.

Chapter 2 presents methods for studying political party programs. Chapter 3 includes a discourse analysis in the category of military security, while Chapter 4 addresses political security. The next chapter (5) discusses Polish political discourse on topics related to economic, social, and environmental security.

The last chapter – 6 – analyses Polish political discourse regarding its securitization, based on the presented research results related to specific identified areas, indicating the dynamics of the process, its nature, and potential desecuritization. The analysis references not only the classic theory of securitization but also critical approaches. It also covers aspects included in the security discourse that are not addressed by securitization processes because they fall within the logic of ordinary politics, and it discusses the rhetorical use of the term security, found, among others, in populist narratives.

CHAPTER 1

Securitization, Riskification, Discourse

No other word has made such a name for itself as *security* today (Kaczmarek, 2008: 16). Classified under the basic categories in social sciences and humanities, it is one of the most widely used concepts in everyday life, resulting primarily from the widespread awareness of both the importance of security and the existence of a large scale of dangers (Stańczyk, 2017: 241). Security is considered to be an "essentially contested concept"; that the meaning of security "is inherently a matter of dispute because no neutral definition is possible" (Smith, 2005: 27). Hence, postmodernists assume that security, as an inferred concept, should be regarded as contentious by its very nature (Taureck, 2006: 55). What an individual understands by this concept "depends on their origin, nationality, education, and political and philosophical views" (Booth, 1997: 118). Therefore, it is natural for individuals and their groups to present different understandings of security and its absence, which in turn can lead to misunderstandings.

Nevertheless, hardly anyone refrains from using it extensively. According to Paul Williams (2012), "every day in the world, people are killed, starved, tortured, kidnapped, threatened, mutilated, reduced to poverty, imprisoned, exiled, and denied education." There is no indication that this situation will change quickly. Hence, Booth's shortest yet accurate definition of security as "survival plus" – survival in freedom from threats to human life plus "other relevant factors" that are formulated depending on who defines that (Booth, 2007) – seems to be a good reference for the observed situation.

Security is one of the key concepts in modern communication, eagerly used by experts, politicians, and the media (especially their youngest form, i.e., social media), as well as by so-called ordinary people. It is used instrumentally in the struggle for the allocation of state resources: it adds weight to the institutions associated with it, and referencing it typically yields excellent results (Buzan, 1991: 370). It has justified various types of reforms, restrictions, and decisions (or the absence thereof), as well as expenditures, at least since it became widely acknowledged that there can be no cutting corners when it comes to security. It also serves as an explanation for withholding specific facts or information from the public, ranging from the identity of soldiers conducting special operations to details of contracts for the purchase of multi-purpose aircraft. To justify disclosing them, it is sufficient to demonstrate that security necessitates it. Securitization, i.e., incorporating a topic into the security sphere, seems to almost always guarantee its elevated status and the opportunity to implement the proposed security scenarios.

Therefore, security, functioning as a meta-operator with significant impact, becomes a driving force; it can assist in addressing problems that, theoretically, are distant from security issues (as seen in the securitization of the AIDS threat and the need to combat the spread of HIV in Africa; Elbe, 2006) or expose its destructive power towards them (securitization of migration issues leading to alarmist perceptions of dangerous immigrants arriving in Europe; Huysmans, 2006: 57). It emerges as an instrumental value (Booth, 2005: 22), "an inter-subjectively constructed secondary phenomenon" (Booth, 1997: 106), but one of significant (and still growing) importance. This initiative to securitize the migration crisis in the middle of the second decade of the 21st century enabled opponents of immigrant reception in several European countries, including Poland, to win elections (or achieve significant success in them), which ultimately resulted in the failure of the EU's migration policy and the enacted plans to relocate refugees. The reasons for this situation lie not only in the nature of securitization itself but also stem from the fact that the category of security is usually created, defined, and classified through a multi-paradigm approach, introducing a variety of security contexts. The absence of a single definition of security implies that in political discourse, anything can be labelled as security, often solely to achieve a rhetorical, populist-driven effect. Thus, labelling something as security does not always indicate that it rightfully belongs to this category, although this category continues to expand. It can manifest as a mere rhetorical procedure, excessive subjectivization, or sometimes even a blatant falsehood.

1 Security Definition and Its Evolution

Definitions of terms, including those of security, reflect the reasoning at a given time and evolve with changes in this reasoning. The concept of security depends on context and time (Rothschild, 1995; Bourne, 2014: 1). For example, Bertrand Russell said that "security as such is a negative goal, inspired by fear" (Russell, 1997: 96), while 10 years later, Ken Booth noted that this category cannot be determined solely by survival because "it only means staying alive, while security means living." In his view, it is a reflection of living conditions that serve human needs (Booth, 2007: 106–107). Time is crucial in defining and understanding security. The introduction of non-state actors into the discussion about security at the end of the 20th century resulted in the debate extending to other levels (above and below the nation-state) (Wendt,

2014). Thus, it broadened the understanding and conceptualization of security. However, the fact that some researchers maintain a division into internal and external issues in the 21st century, which no longer aligns with the contemporary analysis of the security situation in a complex global world (the trap of territorialism), results in limitations on the development of this discipline of science (Bourne, 2014: 21; Morgenthau et al., 2006). Changes in approach over time and the inclusion of risk prevention in the security sphere also mean that today the elimination of uncertainty, rather than threats, and the provision of assurance and control, rather than security, are discussed more often (Derian, 2009; Dillon, 1996). It is therefore difficult to speak of a single understanding and definition of a concept even over short periods, such as one or two decades.

In colloquial terms, security refers to survival, the safeguarding of existence, and freedom from threats posed by others. In scientific terms, "it pertains to the protection of specific goods or values, such as human life, political sovereignty of the state, democracy, identity, territory, property, etc." (Bourne, 2014: 1). It can therefore be stated that it involves the protection of certain goods that have already been acquired. If we attempt to measure them, it is necessary to determine not only an objective level of freedom from threats to these protected goods but also the subjective absence of fear regarding the potential emergence of such threats. However, this is not feasible; hence, one can speak of varying levels of security without assigning a specific value. Furthermore, it is impossible to achieve a status of 100% security (Wolfers, 1952: 484). Therefore, one of the proposals (Baldwin, 1997) to define, describe, and conceptualize security involves answering the following questions: Whose security does it refer to? What goods/values do we seek to protect? To what extent can we provide security? From what threats? At what cost? For how long?

The manifestation of security in various spheres of everyday life, along with the activities of states and international organizations, and the evolving possibilities of defining security, allow for varying definitions based on both modern and historical knowledge. At the same time, this results to various configurations and combinations. Such dynamics influence the diversity of ideas and increasingly complex discourses. Security research has now become a hybrid field, with fundamental assumptions and challenges often arising from broader theoretical and practical endeavors. It has experienced periods of significant theoretical innovation and a focus on empirical issues, resulting in a shift in its scope of interest (Walt, 1991; Booth, 1991; Baldwin, 1997; Prins, 1998; Buzan and Hansen, 2009; Bourne, 2014).

The author's views are of great importance in formulating the definition of security, particularly regarding whether they align with realism/neorealism, liberalism/neoliberalism, or constructivism/post constructivism. This arises from the differing interpretations of social reality through positivist theories such as realism, liberalism, or behaviorism, and postpositivist approaches (including postmodernism, constructivism, and critical theories). Positivism, which focuses on the ontological aspects of security (explaining the world), offers a different perspective than that provided by post positivism (the epistemological approach). Therefore, it is not surprising that the author's views on the potential definition of security influence the understanding of security challenges and threats, as well as the inclusion of various actors, including individuals, social groups, states, and international systems (Zięba, 2012: 13). The inconsistency of changes in the security environment is another barrier to reaching a jointly agreed research position (Glen, 2011).

Often, in situations where it is impossible to specify a definition, referring to the dictionary or the etymological derivation of the concept of security = *sine curas* (*securitas*), i.e., without care, without protection (Kopaliński, 1983: 147), it is understood and explained by its opposite: danger. This results in a significant limitation in comprehension, confined only to the negative sphere (security from …) and a rejection of the trend that focuses on the positive sphere (security to …). This situation is currently widely criticized in various circles, although tolerating ambiguity in concepts is generally considered harmful to science (Huysmans, 1998: 231; Pytkowski, 1985: 59). However, this does not change the actual circumstances related to defining security.

The 1990s and subsequent decades of the 21st century, which brought the opening of the security category to non-traditional military, political, and legal contexts (Zięba, 2008; Stańczyk, 2009), did not escape the "error of over-inclusion" by recognizing that security already applied to almost everything (Frankowski, 2005; Bourne, 2014; Stańczyk, 2011). The concept was undoubtedly at risk of blurring, which would hinder any consensus on its definition among researchers. Despite this, it is difficult to accept today that security can only concern what affects physical survival, even if it is acknowledged that this survival is broadly interpreted, also considering factors such as development or identity (Stańczyk, 2017: 34). A return to a narrow understanding, which could yield a single universally recognized definition of security, would contradict the notion of equality between security and welfare, material and mental comfort of life, good health, social security, certainty of tomorrow, and even mere well-being (Booth, 2007: 104; Lisiecki and Kwiatkowska-Basałaj, 2000).

The lack of a single definition of security, both at the scientific level and in common understanding, as well as the varying definitions of security across disciplines such as law, defense, military science, political science, international relations, and pedagogy (Bourne, 2014: 2; Cieślarczyk, 2009: 9; Brzeziński,

2009: 30), alongside the concept's remarkable rise in popularity, especially in populist narratives that continually expand its subjective scope into new areas affecting individuals and society, complicates its analysis within Polish political discourse. However, when undertaking this analysis, it is essential to seek and describe those definitional areas (and phenomena) that appear to be shared among the broad community engaged in exploring, describing, and creating security.

It is somewhat widely recognized that security is an anthropocentric category (because it focuses on man) (Bourne, 2014: 27; Kukułka, 1982: 29), concerning the human being, which can be described as a value, a common good, and the purpose of human action, framed by his needs in terms of state or process. The individual organizes his or her world of values around it. Anthropocentricity imbues security with subjective and objective dimensions, both of which are extremely important in the practice of ensuring security (Frei, 1977; Zięba, 1999: 28). However, this approach raises the question of whose security it is: that of individuals (human security) or of the collective (society, nation, etc.). Are they equal, or can the needs of individuals be disregarded in the name of collective security? Will the extensive individualization of security approaches and perspectives not lead to the creation of new threats? The strong emphasis on the subjective dimension of security means that, depending on the subject and the situation, in addition to the understanding of security as a value, a need, and the purpose of action, it can also be equated with a product/good or service (provided by public or private entities) (Wendt, 2014; Wolanin, 2005: 12) or with the legal norm regarding orders, rights, and limits of freedom (Hołyst, 2014).

The polysemantic nature of the concept of security (Buzan and Hansen, 2009; Bourne, 2014; Świniarski, 1999: 13) and the absence of a universally accepted definition mean that various attempts to articulate the content and security-related phenomena, which take into account specific criteria, can be found in the relevant literature. If organized chronologically, as they are formulated, approaches that define security through its *alter ego* – threat (considered the primary category in relation to security) would appear first. Stańczyk refers to these as "antonymic definitions" (Stańczyk, 2017: 90). This category includes both the simple dictionary definitions mentioned earlier, referencing the original etymological meaning of the word, as well as more complex descriptions that refer to various threats. In simpler terms, they define what security is not or should not be (Bellany, 1981: 1). They characterize it as a state of non-threat, gradually clarifying that such an ideal state does not exist, thus relating to the absence of existential threats or those we cannot manage. In these definitions, terms such as no threats, freedom from them, elimination, reduction, control, etc., can be found. Given the unlimited number and variety of potential threats, Ficoń suggests that when defining the state of security, one can discuss an acceptable and fully managed level of threats that can be addressed, thereby allowing the security environment to be consciously shaped (Ficoń, 2013: 11). This approach reinforces the role of the state as an institution that can and should act accordingly.

The simplicity of antonymous definitions may be their advantage, but it is challenging to use them to describe security in a post-modern global world. The lack of threats (even existential ones that cannot be prevented) no longer equates to security and, more precisely, does not cultivate a sense of security in individuals and their collectives. We already know that a threat does not have to diminish the level of security, which has been traditionally characterized by military perspectives that emphasize force and power (e.g. Waltz, 1959). Military force ceases to play a crucial role in ensuring security when economic, political, and even geographical indicators determine the supremacy of states. Therefore, there is a need for broader approaches in defining security. The first step on this path was to expand the negative definitions (freedom from threats) with positive "threads." In the mentioned typology of methods, Stańczyk defines them as horizontal, complementary definitions (Stańczyk, 2017: 99), pointing out that the "hard" elements (pillars) of security, such as armed forces, are augmented with "soft" elements that consider the redefinition of the concept of force. The concept of positive security appears here, which is associated not only with the ability to respond to an expanding array of challenges and threats but also with leveraging security opportunities (Zięba, 2012: 13). Another approach is to try to determine what security is, depending on the specific context, by referencing other conceptual categories (Stańczyk, 2017: 103). Security, in this context, becomes a need, a goal, a value, an interest, a commodity (a rare, expensive good), a service, and also a mission or system (Jakubczak and Flis, 2006: 20). It can also be represented as certainty, survival, welfare, and freedom. Because these terms are already defined, it becomes easier to define security using these categories.

This approach also includes the division made by some researchers into security understood as a state and process (Huysmans, 1998; Morgenthau et al., 2006; Ficoń, 2013: 10; Antoszewski, 1998: 176; Baldwin, 1997; Prins, 1998; Buzan and Hansen, 2009; Bourne, 2014; Mikiewicz, 2005: 35; Stefanowicz, 1984: 18; Brzeziński, 2009: 30). When defining security as a state, it should be noted that the situation in which the subject is located is not constant (so it can only be described "here and now"). Contextual definitions (as Stańczyk calls them) also consider security in ethnic (national security) and state terms (state security). A category of international security emerges as a result of the intersection of the security of different countries (Stańczyk, 2017: 106). This subjective dimension is extremely important when defining security in modern times. Different approaches, depending on whether they concern the security of the state (a trend popular in definitions created within the framework of international relations) or of individuals and humanity (postmodern approaches), allow for a distinction between personal and structural security (Drabik, 2013; Rosa et al., 2007; Świniarski, 1997). Post-modern approaches to defining security, particularly those developed after 2001, account for the increased roles of non-state, non-territorial, and non-institutional actors in international relations, which also influence security dynamics. Concepts of social, environmental, economic, and human security will emerge, as further discussed in the next chapter. This will refer not only to individuals and their security but also to social groups living in specific security environments (e.g., immigrants). At the state level, national security remains a focus, primarily achieved through military capability and external politics, as well as in internal contexts (public and universal security contexts; Ścibiorek et al., 2017). Political decision-makers must view security challenges from both institutional and social perspectives, which will affect political discourse. The international level of security, operating above the state, encompasses both regional (regional security complexes; Buzan and Wæver, 2003: 44) and global dimensions. Definitions formulated from these perspectives are referred to as vertical by Stańczyk.

Another approach to defining security within security studies, which seeks to address the ambiguity of the concept, is the use of classification. Subjective and objective, as well as processual typologies, are most often cited here (Wolfers, 1952; Zięba, 1999; Pietraś, 2007; Wojtaszczyk, 2002; Jakubczak and Flis, 2006; Baldwin, 1997; Prins, 1998; Buzan and Hansen, 2009; Bourne, 2014). Therefore, on one hand, it pertains to the security of the individual, group, society, nation, state, international community, and even humanity. On the other hand, typologies are largely shaped based on threats, the source of which may be external or internal to the subject and may involve military, economic, health, social, or political issues, etc. These threats can be symmetrical or asymmetrical (the comparison of the potentials of entities) and may pose a threat of varying scope (from individual to global), stemming from an unexpected result of a change in the system of forces or the emergence of new destabilizing factors (Żurkowska, 2007: 36–37). The universality of the category of security is evident in that it involves various personal and structural entities at sub-national, national, and international levels of analysis, and today, in a specific way, at the transnational level (Kitler, 2010: 18). In the processual dimension indicated by Kukułka (1994: 34), security is understood not only as a state but also as a process, which leads to the need to consider its dynamics resulting from changes in ways of strengthening security, the interdependence of the security interests of modern states, and the inseparability of international security and peace (Zięba, 2006: 939). In this approach, the boundaries between the internal and international dimensions are blurred, and the complexity of this new security is reflected in the name proposed by Fischer (1993: 13), namely comprehensive security.

For the purposes of the discussion and research conducted in this monograph, it was assumed, following constructivists, particularly B. Buzan and the Copenhagen School, that security is no longer only a matter for states (they are not its only subjects) and is not limited solely to the use of armed forces. It is multidimensional, more extensive than Buzan's five categories, and is a social construct (Fierke, 2007: 56), the content of which depends on the context. Security consists of striving for freedom from threats and the ability of states and societies to preserve their independent identity and functional integrity against the forces they perceive as hostile. Survival is the basis for security, but it also encompasses issues related to the quality of living conditions (Buzan, 1991: 432).

The constructivist approach underlying securitization is considered appropriate for analyzing the political discourse on security, given the role of language in it (the way in which security is talked about and thus constructed). Some security issues are original in relation to language (they existed before being termed as such, usually because of their objective nature), while others were categorized as security only after this interpretation (Kostecki, 2012: 115). In the discourse, including the political one, both references to "old," objective security problems/issues and new ones recently classified as such (known as security issues) can be observed, as well as the inclusion of further areas in this category. Political discourse is also characterized by the selective (subjective) treatment of objective security problems and their subjective and intersubjective prioritization.

2 From Realism to Constructivism: the Copenhagen School and the End of the Cold War: Security as a New Black

The two main theories of international relations (and therefore the two main classical approaches to understanding security) – realism and liberalism – perceive security as the domain of states and regard the principles of sovereignty and non-interference in the internal affairs of other states as the basic guarantees of state security. According to these theories, the best way to achieve security is to build the foundations of international order (Bellamy,

2009; Beaulac, 2004). The military power of states is intended to maintain this order, both at the level of deterrence and against existential threats.

These theories seem sufficient to explain and provide security, as long as one assumes that states are the best guarantors of human security. However, "in the 20th century alone, 260 million people were killed by their governments – six times as many as were killed in combat by the armies of other states during the same period" (Rummel, 1994: 6), including in two world wars. Thus, while proponents of the traditional study of security continue to view the world as state-centric, even statistically (Booth, 1991: 320), recognizing that the proper subject of security and the factor providing it is the state, which is the main center of decision-making and social loyalty (Bilgin and Morton, 2002), the circumstances with which they collide demonstrate the anachronism of such an approach and the impossibility of explaining through it the events and processes occurring in the modern world.

It should be remembered that the traditional realist approach links security to matters of utmost importance, such as war, death, or survival. For years, crime, economic, or pandemic (health) security was outside the interest of international relations scholars and remained primarily within the realm of strategic studies. Aspects of security and public order, which we analyze today, were classified even lower, as part of the domain of "law and order." During the Cold War, the constant threat from the Warsaw Pact, Reagan's Star Wars, and a bipolar world made such an approach – focusing on war and physical survival – seem reasonable, even though it faced strong criticism from constructivists, particularly from the Copenhagen School, as early as the 1980s. However, it was only after the collapse of the bipolar world and the identification of terrorism and organized crime as new global threats that security studies embraced a perspective acknowledging that collective security is essentially the sum of individual security (human security) and encompasses not only military or political dimensions but also economic, social, and environmental aspects (Table 1), as Barry Buzan pointed out a few years before the fall of communism in his famous People, States, and Fear (1983). The founder of the Copenhagen School argued that the state does not have a monopoly on security in its military, political, economic, environmental, and social sectors. It controls the first two zones and sometimes even the third, but the situation in the last two sectors is more complex (Bigo, 2012: 117) as states lose in this field to international and cross-border institutions or global corporations. Thus, the combination of post-Cold War optimism and the receding of the persistent military threat created space to consider non-military issues that posed a security threat (Vogler, 2013: 19).

TABLE 1 Security categories according to B. Buzan

Category	Definition	Key value
Military security	Concerns the two-level interplay between states' offensive and defense capabilities and their perceptions of each other's intentions	sovereignity
Political security	Concerns the organizational stability of states, their systems of governance, and the ideologies that provide them with legitimacy and power	survival
Economical security	Concerns the level of a state's access to the resources, finances and markets necessary to maintain an acceptable level of prosperity and state power	prosperity/ welfare/ well-being
Societal security	Concerns the durability of traditional patterns of language, culture, religion, national identity and customs	identity
Environmental security	It is concerned with maintaining the local and planetary biosphere as the fundamental support system on which all other human endeavors depend	sustainable development

SOURCE: BUZAN (1991: 19–20)

Steven Walt, David Baldwin, and others outlined the evolution of mainstream security studies over four decades leading up to the mid-1980s (Walt, 1991; Kolodziej, 1992; Baldwin, 1995). Security studies scholarships have focused on topics such as weapons proliferation, nuclear deterrence theory, military strategy in counter-insurgency wars, arms control, the security dilemma, alliance formation and dynamics, the offence–defense balance, and more (for examples, see Snyder, 2007; Shiping, 2010; Williams and Viotti, 2012; Krause and Williams, 1997). As Baldwin put it, security studies "has tended to focus on one set of means by which security may be pursued," and has "tended to assert the primacy of military security over other goals" (Baldwin, 1995: 129), missing many of the most intense dynamics of contemporary security relations: in the areas of national identity and culture; concerning the security of groups, individuals, or the biosphere; regarding questions of state stability or fragility; involving economic dislocation; and concerning global flows of people and information; as well as with structures and institutions of security cooperation (Krause and Williams, 1996).

As Krause and Williams (1997) point out, this inability to capture contemporary security dynamics has necessitated broadening and deepening the security studies agenda. By broadening, they understand addressing "threats and sources of insecurity beyond the military security of the territorial state, the most prominent of which have been environmental or economic challenges, but which have also included issues such as transnational migration, global health, food, energy, or human rights." Deepening means including different referent objects, such as individuals, sub-state groups, states, regions, the global system, and the biosphere, to capture the complexity of contemporary security dynamics and the political and ethical issues involved in studying and practicing security. It also involves moving away from an exclusive focus on the state toward individuals or social groups below the level of the state (societal or human security) (Buzan et al., 1998; Hanlon and Christie, 2016) or toward institutions and structures above it (regional security arrangements or wider cooperative security mechanisms) (Buzan and Wæver, 2003; Neack, 2017). A further extension of the broadening/deepening argument proposes introducing a positive understanding of security, not only through the threat perspective.

To prevent security emancipation (Booth, 1991) from extending excessively and unjustifiably, and following Baldwin (1997) while considering the classic paper by Wolfers (1952), it is essential to enhance and broaden our understanding of security by addressing these questions: "security for whom?, security for which values?, security from what threats?, security by what means?, security in what period of time? And finally: how much security?"

Critical Security Studies (CSS) emerged after the Cold War as critical interpretations of preexisting notions of state-centricity and conventional claims that state sovereignty equates to security (Robinson, 2010; Wallace and Romaniuk, Krause and Williams, 2018: 257). Although there is no universally accepted definition for CSS, Booth (2005) defines it as a theoretical commitment that involves a critical and ongoing exploration of ontology, epistemology, and the praxis of security to enhance security through emancipatory politics. Conversely, Buzan et al. (1998) assert that CSS focuses on individuals and the deconstruction of previous security constructs. Bilgin (1999) attempts to operationalize CSS, pointing out that it favors an explicitly normative security agenda based on human emancipation, in contrast to Cold War security agendas that, under the guise of objective theories, prioritized the security of states above all else (Wallace and Romaniuk, Krause and Williams, 2018: 257–258).

The Palgrave Encyclopedia of Global Security Studies (2023) indicates two major CSS schools: the Copenhagen School and the Aberystwyth (Welsh)

School. The first will be described in the following sections of the book. The Welsh School operates within the tradition of critical theory, which has its roots in Marxism. Diskaya (2013) points out that CSS is based on the pioneering work of Ken Booth and Richard Wyn Jones, who were heavily influenced by Gramscian critical theory, the Frankfurt School critical social theory, and radical international relations theory. Among representatives of Critical Security Studies, one can identify The Paris School of Security Studies (concept of insecuritization), Feminist Security Studies (Feminist Security Theory), post-structuralist theories related to identity construction (such as R. L. Doty, 1993), and reflexive security studies (in general), which offer sociological insights into the 'risk society' (see Vedby Rasmussen, 2004). Additionally, The Minnesota School (R. Duvall) will be characterized below in the section about critics of The Securitization Theory.

Paradoxically, the collapse of the bipolar world did not herald an immediate increase in the significance of security issues, either internationally or locally. Societies on both sides of the Iron Curtain anticipated a respite from high military spending, a reduction in armies and their maintenance costs, and an end to discussions about security as they knew it. The free market and liberal democracy were expected to ensure peace, respect for human rights, and deter aggression in relations between states. Security, as traditionally understood (the military dimension, the domain of states), was indeed losing its importance. This allowed other approaches to emerge fore.

Going beyond the military dimension in analyzing security assumes that security is a social construct (Fierke, 2007: 56), shaped by various factors and circumstances and interpreted and understood differently. Something becomes security (or a threat) not (or: not exclusively) because objective factors determine it. The example of the 2003 invasion of Iraq by the US and its allies can be cited in this context, when these countries considered the possibility that Saddam Hussein possessed nuclear weapons to be more dangerous than the actual nuclear capabilities of countries such as Russia, China, North Korea, France, India, Pakistan and Israel. It is clear that the reasons for recognizing Hussein as a threat are historical (the first Gulf War), political (the Iraqi regime), and cultural (the opinion of the pariah state in the international community) aspects (McDonald, 2012: 61–62), rather than a comparison of the nuclear potentials and ambitions of the mentioned countries in this regard. Security is thus constructed in politics and is closely dependent on accepted social norms, which are not given once and for all. How they change affects the recognition of what constitutes a threat and how to mitigate it. At the heart of the constructivist approach to security is the idea that threats are called into existence, rather than simply stated by comparison with some abstract set of criteria that remains unchanged over the years. In other words, the catalogue of areas identified as security-related will change as new issues are identified or old ones are excluded, as long as there is an effective act of securitization.

Growing out of a constructivist foundation, the Copenhagen School (Buzan et al.) has progressed further in evaluating the significance of context and conceptualizing the creation of security problems. Capitalizing on a climate favorable for broadening the definition of security, it has transformed issues such as environmental change, poverty, and human rights into crucial axes of security discussions as viewed from the perspective of individuals or social groups. The success in promoting this approach can be attributed to the careful preservation of analytical rigor in expanding the scope of the dialogue regarding what security currently encompasses and what it does not yet include, particularly highlighting the division of security into sectors (military, political, economic, social, ecological), within which various forms of relations among the actors involved in the formation of security occur (Buzan, 1991).

For Buzan and his colleagues, nothing was surprising in the fact that in each sector, the subject and object of security were defined differently, and different solutions were proposed for ensuring it. In this regularity, they saw fertile ground for the development of the theory of securitization, explaining the "success" of the concept of security in the modern world.

In the objective sense, security is the state of being free from threats, while in the subjective sense, it is the absence of fear or the fear that these values will be threatened (Wolfers, 1952: 485). Daniel Frei (1977: 17–21) identifies four states resulting from this distinction:

1. the state of insecurity, characterized by the presence of a large actual external threat and the correct perception of this threat (adequate);
2. the state of obsession, which occurs when an insignificant threat is perceived as a large one;
3. the state of false security, which refers to the situation when the external threat is serious, but it is perceived as slight;
4. a state of security occurring when the external threat is insignificant and its perception is correct (Zięba, 2012: 11).

Discursive accounts of security, which will serve as the foundation for securitization theory, assert that it is impossible to define security objectively. For these accounts, it is crucial to understand the process through which specific threats manifest as security problems. In other words, threats become "objective" when they are recognized as such, not because they possess some innate "characteristic" (Buzan and Hansen, 2009: 32–33). Researchers from the Copenhagen School describe securitization as the process of transforming challenges into threat categories (Zięba, 2012: 13). This was evident, for instance, in the assessment of environmental degradation, which, back in the 1980s, was regarded as a challenge but was later recognized by politicians and societies as a threat. Simply put, it can be assumed that challenges and threats exist along a continuum, where a need gradually arises and ultimately transforms into the necessity for an entity to take action to ensure its security (Zięba, 2012: 13).

Securitization Theory posits that for threats to be considered security issues, they must meet a specific set of criteria "that distinguish them from the normal course of exclusively political issues" (Hough, 2014: 23). Buzan, Wæver, and de Wilde expanded the spectrum of potential threats to include non-military issues, maintaining that such issues can be considered security issues even if they do not impact military or commercial concerns (Hough, 2014: 23; Dourley, 2016: 2).

The key criterion for securitization according to the Copenhagen School is that the existential threat to the referent object must be recognized and treated as a matter of exceptional politics implemented outside the normal democratic process (Oels, 2012: 185). Securitization refers to the category of emergency politics, which are "urgent and immediate situations that normal political life cannot resolve" (Floyd, 2012: 287). The process of raising or transforming a security threat is conducted through performative "speech acts" that highlight the danger of the problem, elevate its political profile, and justify the need for exceptional measures (Vogler, 2013: 19). In Phase I of this process (identification), an issue is identified as a threat, and in Phase II (mobilization), a request for extraordinary action is made (Roe, 2008) (Figure 1). This act involves two key actors: the securitization elite, who are responsible for presenting the problem in securitized terms, and the audience, who justify the securitization of the threat and the need for exceptional measures (Oels, 2012: 191). For example, a massive bailout package that uses taxpayer money to save a failing bank can be justified by politicians who tout the necessity of such measures to save the financial system (Oels, 2012: 191). It is hard to disagree that the scenario outlined is clearly outside the regular processes of the democratic system and can probably only be implemented if a sufficiently representative portion of the population accepts the speech act as justification for emergency action.

There are three principal components in securitization theory: (1) the referent object: what is perceived to be existentially threatened and has a legitimate claim to survive; (2) the securitizing actor: an entity that securitizes issues by declaring something (a referent object) existentially threatened; and (3) the audience: the target that must be persuaded that the referent object is existentially threatened (Buzan, Wæver, and de Wilde, 1998). In the process of

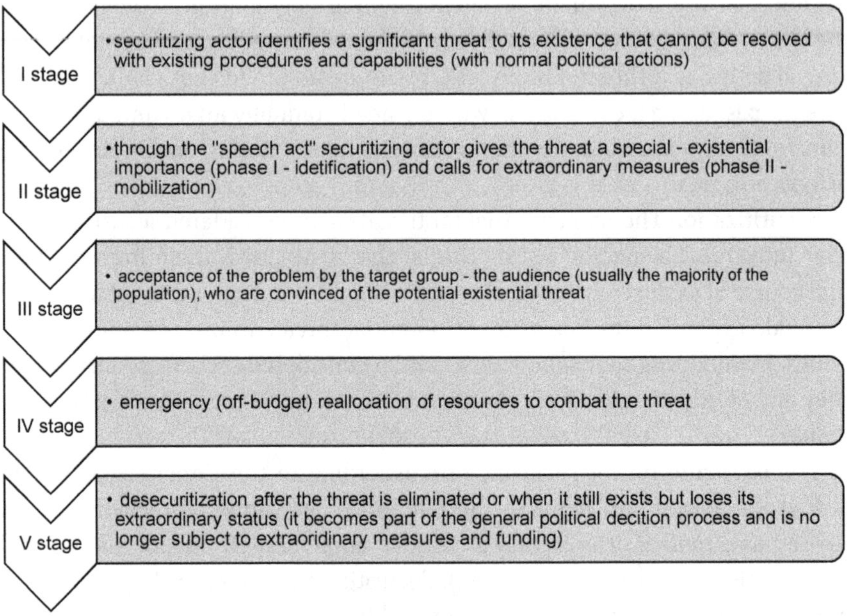

FIGURE 1 Securitization and desecuritization stages
SOURCE: OWN STUDY

securitization, security becomes a subject of negotiation between the securitizing actor presenting the threat and the audience. One can only speak of the effectiveness of securitization if this narrative and the extraordinary nature of the situation as it relates to security (most often articulated by political elites) are accepted by the "audience." Security categories are thus created here by a "speech act," most often employing a specialized language of security and threats. They enable a rapid transformation of issues previously unrelated to security (such as immigration, which traditional security studies do not recognize as relevant) into security-critical topics. In the US and Europe, following the wave of attacks in the 21st century, especially post-9/11, migrants were officially classified as a security threat, prompting formal responses like closing borders and even using military force (Huysmans, 2006). Successful securitization can be identified when the normal "rules of the game" are suspended in favor of extraordinary measures (Buzan et al., 1998: 34). Although there are instances where social acceptance is not necessary, the existential nature and suddenness of the threat justify extraordinary actions. However, it must be acknowledged that securitization theory is far from precise in its depictions of the interaction between the securitizing actor and the audience, as well as in the necessity for acceptance of the perceived threat and

TABLE 2 Practical aspects of securitization according to the Copenhagen School

Securitizing actor	State, civil society, international and local non-governmental organizations, UN agencies, individuals (environmental activists), governments and their military elites, political leaders, lobbies
Audience	Public opinion, politicians, military elites, society, government and non-government actors
Referent object	Economic and political stability, identity, sovereignty, territorial integrity
Speech act	Statements, declarations, documents, media coverage, materials in both verbal and non-verbal form.
Securitization movement	Describing an existential threat to important values
Securitization act	Recipients accept a given speech act by accepting the claim of influential political elites that an existential threat exists and then approve a response using emergency measures.

SOURCE: OWN STUDY, BASED ON: MUSIOŁ (2010: 49)

the adoption of extraordinary measures (Fijałkowski and Jarząbek, 2016: 32) (Table 2). Even aside from this inaccuracy, one can still question, as Fijałkowski (2013: 95) rightly suggests, whether the success of securitization is genuinely a desired outcome, given that the very occurrence of the process is a sign of policy failure.

As Balzacq argues, "it is useful to think of security pronouncements not as speech acts that are successful to the extent that rules are followed by the agents but as discursive techniques allowing the securitizing actor to 'induce or increase the [public] mind's adherence to the thesis presented to its assent'" (2005: 172). In this model, securitization is a "sustained strategic practice aimed at convincing a target audience to accept, based on what it knows about the world, the claim that a specific development (oral threat or event) is threatening enough to deserve an immediate policy to alleviate it" (Balzacq, 2005: 173).

The securitizing actor securitizes issues by declaring that the referent object is existentially threatened by an outside source. To perform the role of a securitizing actor, this element must be perceived as an epistemic authority that impacts the audience and convinces it that a certain issue constitutes a security threat to the referent object (Michael, 2009). While the actors that commonly fill this role include political leaders, governments (Buzan et al., 1998, p. 40), and the military in cases like Israel (Michael, 2009), other actors such as the media (McDonald, 2008), NGOs, and environmental groups (Trombetta, 2011) can also raise an issue as a security threat.

As Salter (2008: 330) and Leonard and Kaunert (2011) point out, there can be a large number of audiences, each characterized by specific participants. However, they do not explain why each specific audience's support is essential for securitization to occur, while the acceptance of others is not vital for the process. Moreover, in reality, the securitizing actor cannot deal with numerous audiences; therefore, he or she must "focus on those necessary to be convinced for conducting securitization" (Wertman and Kaunert, 2022: 71–72).

The audience is not merely an element that authorizes or rejects the securitizing actor's securitization move. In fact, the audience is an active agent capable of having a meaningful effect on the intersubjective construction of security values. Thus, "the audience also has the ability to influence securitization processes and the policies selected to address perceived threats" (Cote, 2016).

In this book, we will understand the securitizing actor as political decision-makers (political parties and political leaders, acting and pretending to act) and the audience as the society entitled to vote in elections, as well as the political majority when extraordinary measures need to be voted on.

Defining something as a security issue makes it such an issue, Ole Wæver (2004) argued. At the same time, he believed that the process of securitization is a failure because it indicates that ordinary politics cannot solve the problem and requires an intersubjective construction of existential threats (which calls for extraordinary measures) under the threat of the annihilation of the state or society (a threat to sovereignty or identity). This phenomenon shifts from the realm of ordinary politics to the realm of existential threats of "highest intensity" (Buzan and Wæver, 2004). Securitization threatens to "dedemocratize," depoliticize, generate conflict, and increase the security dilemma (Floyd, 2016a). Wæver believed that only desecuritization, which occurs when extraordinary measures cease and the securitized issue returns to the realm of normal politics, can be regarded as a positive phenomenon. Securitization is the opposite of politicization, which involves incorporating the broadest possible spectrum of issues into public debate. Thus, in this context, politicization is a positive phenomenon, while securitization, which excludes specific issues from public debate, is negative one.

The process of desecuritization begins when the fear index of a threat is lowered due to taking protective and/or hazard-avoidance measures. For example, the threat of an energy supply disruption is mitigated by signing energy supply contracts with alternative trading partners (energy suppliers) (Czaputowicz, 2007: 191). Buzan et al. (1998) define desecuritization as a process where issues are shifted from emergency mode to the normal political process. For Wæver, as previously indicated, desecuritization is preferred over securitization (Wæver 1998). Thus, there is a moral-legal problematique surrounding securitization that desecuritization addresses. Much debate exists about how issues may be returned to the normal political process. For Wæver (1995), this may occur in three ways: first, by removing the securitization label and status from an issue by refraining from discussing it in security terms. Second, it may occur by limiting security dilemmas through curtailing security responses and, finally, by returning the issues to everyday politics through communication. However, Behnke (2006) challenges Wæver's idea on this point by arguing instead that with desecuritization, an issue will disappear through silence, not by using a speech act to revert it to normal politics (Haughton, 2023: 1299).

As Rita Floyd (2007: 330–335) and Claudia Aradau (2004: 389–390) note, the concept is 'largely under-theorized and open to interpretation'; it is undermined analytically or politically by an 'insufficient attention to politics in the theory of securitization,' and there are no normative connotations to the Copenhagen School due to its repudiation of the concept of emancipation (Hansen, 2012: 527). Hansen also points out that the concept is applied in ways that, "when compared, seem unsystematic or even contradictory" (Hansen, 2012: 528). At the same time, she claims that "the choice between securitization and desecuritization must, considering the political status that both concepts have, be inherently political. As a consequence, we cannot derive decisions on whether to (de)securitize from any objective status that threats might have" (Hansen, 2012: 535).

Wæver theorizes three strategies of desecuritization: not discussing an issue as a threat at all, managing securitization so that it does not spiral, and moving the securitized issue back into normal politics (Wæver, 2001: 253). Hansen proposes four deeper forms of desecuritization: (1) *change through stabilization* (this form of desecuritization implies a rather slow move away from an explicit security discourse, which in turn facilitates a less militaristic, less violent, and hence more genuinely political form of engagement), (2) *replacement* (where it is possible to indicate the combination of one issue moving out of security while another is simultaneously securitized), (3) *rearticulation* (an issue is removed from the securitized by actively offering a political solution to the threats, dangers, and grievances in question), and (4) *silencing* (when an issue disappears or fails to register in security discourse) (Hansen, 2012: 539–545).

Securitization processes are intensifying in the 21st century. The approach of analyzing areas such as health, finance, the economy, or the environment through the lens of security is becoming commonplace, and risk management practices are regularly employed to manage vulnerabilities in these areas.

These areas include the epidemic, controlling stock market volatility, recording, controlling, and profiling individuals' behavior in the provision of public security and migration, and managing climate change scenarios (Hardy and McGuire, 2016: 83).

The change brought to the understanding of security by the Copenhagen School (and other representatives of Critical Security Studies) entails a broadening of the catalogue of threats differentiated by the diversity of reference actors. Since threats of a military nature are losing their previous primacy, it is possible to see hitherto minimized non-military factors shaping security in the 21st century (Alker, 2005: 195). Adopt other scenarios of conduct or make different decisions than hitherto, guided by non-military considerations. Especially since, as Garrett (2005: 15) predicts, a greater challenge of the near future will be health care personnel shortages (5 million vacancies worldwide) than military recruitment.

Almost half a century after the first publication of B. Buzan, it is apparent that the approach proposed by the Copenhagen School has no shortage of gaps or unanswered questions. First and foremost, they do not address whether the securitized issues become such solely because of the speech act, or whether there are objective (real threat) factors behind considering them as security issues (Księżopolski, 2011). Thus, it is not entirely clear what determines the recognition (or not) of an issue as a security threat. If the aforementioned "objective factors" are present, it politicizes rather than securitizes the issue in question. It is difficult to make such a distinction based on the Buzan et al. approach. Meanwhile, examining environmental issues, many authors point out that states recognize them as a security threat and respond to them, but without taking extraordinary actions, which are a condition for effective securitization (Księżopolski, 2009; Williams, 2003). Thus, they are politicized rather than securitized (Figure 2).

The reason for these doubts is the political practice of leveraging security issues to strengthen political positions, particularly to achieve electoral success. In principle, as long as this process is successful, it can be considered securitization (speech act, acceptance, extraordinary measures). However, it is essential to note that this can result in erroneous and harmful decisions as a consequence (this will be discussed in more detail in the following pages). As Campbell explains (1992: 1–2), this is due to the Copenhagen School's postmodern assumption of the absence of objective dangers. Thus, the securitization of virtually any area is legitimate, as long as there is a performer of the speech act and an audience willing to support it, who will accept the extraordinary means of action. For some researchers, this approach is not sufficient; they

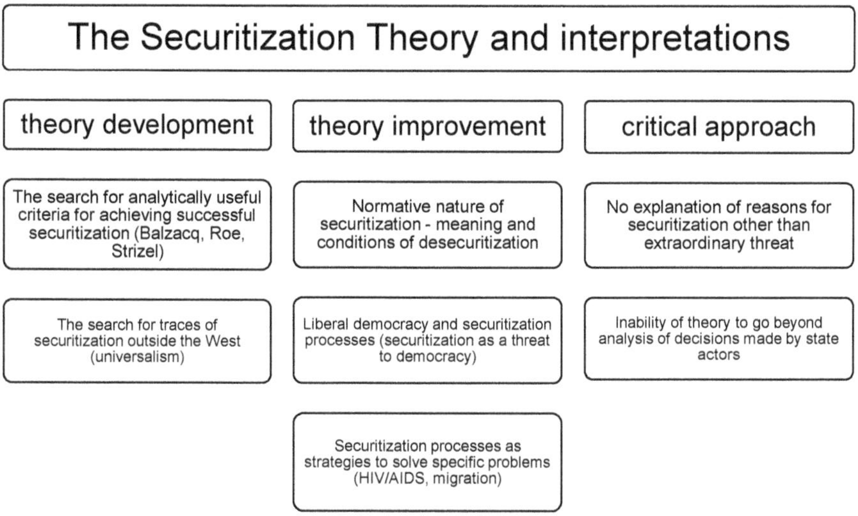

FIGURE 2 Development of the Securitization Theory (Balzacq, 2005; Roe, 2004, 2008; Stritzel, 2007; Vuori, 2008; Caballero-Anthony i in., 2006; Aradau, 2004, 2006; Elbe, 2006; Huysmans, 2006; Williams, 2003; McDonald, 2008; Taureck, 2006)
SOURCE: OWN STUDY BASED ON: FIJAŁKOWSKI (2012: 149); GAD AND PETERSEN (2011: 315–328).

advocate searching for objective factors supported by complex empirical data to categorize a given threat as security (Cha, 2002). The lack of control over what should be securitized and what should not, stemming from the disinterest of securitization theory and its authors in the causes of securitization and the intentions of securitizing actors, can, as Roe (2012) notes, serve as a strong temptation for rulers to utilise securitization to introduce undemocratic laws and strengthen their power. Today, a significant portion of securitization acts results from actions taken by security management professionals and public and private risk management agencies. Many securitization movements are so integral to these mechanisms that they go undiscussed, presented not as exceptions but as something that flows naturally from ordinary procedures, falling within the logic of freedom, which Bigo indicates as a dangerous process for democracy and civil society (Bigo, 2012: 125).

The popularity of securitization theory has led to criticism from many directions. Feminist research has questioned the explanations of gender relations (Hansen, 2000; Heck and Schlag, 2013; Hoogensen and Rottem, 2004; H. Hudson, 2005; N. F. Hudson, 2009; Kearns, 2017). Difficulties in applying the theory in countries outside Western civilization have raised concerns about Eurocentrism, which either hampers this application or necessitates modification of the theory (Bilgin, 2010; Bilgin, 2011; Wilkinson, 2007; Sabaratnam, 2013). Howell and Richter-Montpetit (2020: 115) argued that classical securitization theory is rooted in racism. In fact, this was not the first work to incorporate racial issues into securitization theory (Amin-Khan, 2012; Ibrahim, 2005; Mofette and Vadasaria, 2016). However, the charge made by these authors carries more weight: it consolidates all previous accusations of promoting a European (or Euro-Atlantic) understanding of security by white males who reach for concepts like civilizationism to sustain a merely valid (though exclusionary) understanding of security, its categories, and especially existential threats. According to Howell and Richter-Montpetit, securitization is nothing more than "a potential regression to a racially coded uncivilized 'state of nature'" (Howell and Richter-Montpetit, 2020: 117).

One of the more interesting critiques of the Copenhagen School, useful from the perspective of the research proposed in this book, is the work of Rita Floyd (2007, 2010, 2011, 2016ab, 2019). She not only pointed out the gaps in classical Securitization Theory but also proposed her normative approach, which ultimately resulted in Just Securitization Theory (JST). Her book, aptly titled Morality of Security (2019), organizes earlier considerations about the frailty of traditional securitization theory and proposes her normative approach embodied in the theory of morally justified securitization. The main differences with the theory proposed by the Copenhagen School primarily concern the existential threat itself, which, according to Floyd, should be objective (recognized as such, among other things, through studies of the sincerity of potential aggressors). Floyd refers to the observations of another critic of Buzan, Therry Balzacq, who noted that while it is difficult to identify objective security threats, objective existential threats can already be successfully enumerated (Balzacq, 2005). Secondly, according to Floyd, it does not matter whether the recipients of the speech act – the securitization movement describing this existential threat (Floyd, 2016a) – accept it or not (which was crucial for the Copenhagen School), since the essence lies in action: the practice of security and the implementation of specific policies, rather than merely the acceptance of their description (Floyd, 2010). Figuratively, this can be represented by the equation:

tST: SECURITIZATION = SECURITIZATION MOVEMENT + AUDIENCE ACCEPTANCE;
where: the securitization movement should be understood as a justification for an existential threat
tJST: SECURITIZATION = SECURITIZATION MOVEMENT + SECURITY PRACTICE

According to Floyd, securitization occurs not "when the audience accepts the justification of the existential threat, but when instead there is a change in behavior by the subject, which is justified by that subject using reference to the declared threat. Securitization becomes successful by the fact that it has occurred, without the need to break normally applicable rules or introduce extraordinary measures" (Floyd, 2016b). Securitization is successful only if the identification of the threat justifying the securitization move is followed by a change in behavior (action) by the securitizing actor (or someone else at his behest) and if the action taken is justified by the securitizing actor's reference to the threat identified and declared in the securitization movement. The ultimate object of reference is the human being, and security is not so much (not only) survival as it is the possibility of development (well-being).

Finally, according to Floyd, it is not even necessary to use extraordinary measures when dealing with securitized issues. The "standard emergency measures" enshrined in the constitutions of liberal democracies are sufficient; that is, introducing new laws in accordance with existing procedures; introducing new powers to manage the emergency situation within the framework of the existing legal order, approved by the relevant courts; or, finally, utilizing the existing security apparatus and current emergency legislation to resolve issues not previously addressed. Hence, Floyd's proposed catalog of conditions for just securitization in Table 3.

Floyd advocates not only for changes in the approach to the act of securitization itself but also insists on a clear and prompt termination of securitization (desecuritization) and a ban on securitizing the same problems again. An analysis of practical examples shows that many issues could already be desecuritized because actions to combat the threat are already firmly established in "normal politics" (e.g., HIV/AIDS issues, which, according to the UN, already attract enough attention to not need to be presented as a security issue; McInnes and Rushton, 2013).

While criticizing the Copenhagen School, Floyd does not undermine the very idea of presenting selected problems as security issues. There are situations in which a problem – one that objectively poses an existential threat and is therefore morally "right to be securitized" – needs to be viewed from such a perspective to persuade society to allocate adequate resources to combat it or to compel policymakers to address it when they are trying to avoid it, she argues. However, looking back over nearly 50 years of securitizing various issues since the first descriptions of securitization theory, it is clear that it has sometimes been abused (such as in the case of the 2003 invasion of Iraq), and issues continue to be considered an existential threat requiring extraordinary measures to combat them, even when they have not been necessary

TABLE 3 11 criteria of Just Securitization

Just resort to securitization
1. An objective existential threat to a referent object.
2. Referent objects are entitled to defend themselves or are eligible for defensive assistance if they are morally justifiable.
3. The right intention for securitization
4. The good gained from securitization must be judged greater than the harm securitization is expected to entails.
5. Securitization should not lead to more insecurity than it aims to solve.
Justice during securitization
6. The response must be appropriate and should aim to address the objective existential threat that occasions securitization only.
7. The response must cause, or risk, the least amount of harm possible and at the same time do less harm than there would be if securitization were abandoned.
8. Offenders and suspects detained as part of securitization must be treated humanely
Justice after securitization
9. Timing: desecuritization of just securitization must occur when objective existential threats have been neutralized, whereas desecuritization of unjust securitization must occur immediately.
10. Action: security language and security practice should be terminated in full and with immediate effect
11. Long-term aim: desecuritizing actors should avoid renewed and/or reactional securitization by building a stable desecuritized state of affairs.

SOURCE: FLOYD (2014: 122–125)

for a long time (part of the migration policy challenge). Hence, the author demands an examination of the objectivity of the threat and the intentions of decision makers, and when securitization has already occurred, to desecuritize as quickly as possible and prohibit re-securitization of the same issues.

Table 4 presents a comparison of the most important aspects of the Securitization Theory and Just Securitization Theory.

TABLE 4 Comparison of the Securitization Theory and Just Securitization Theory

	Securitization Theory The Copenhagen School Barry Buzan et al.	**Just Securitization Theory Rita Floyd**
Existential threat	Its occurrence is crucial to the act of securitization	Its occurrence is crucial for the act of securitization under certain conditions (see below)
The objectivity of existential threat	Threats become objective when they are recognized as such, they do not have to be so by their nature	Existential threats have to be objective
Securitization movement	Speech act – justification of existential threat	Speech act – justification of existential threat
The importance of the speech act	An issue becomes a security issue by simply naming it	Constitute warnings (against the aggressor) and promises to defend the reference security object
Audience accepting the speech act	Key to successful securitization	It is not essential; it is important that steps are taken to ensure security
Extraordinary measures	Are the result of transferring the securitized phenomenon from the realm of ordinary politics to the extraordinary politics	Not essential, can be replaced by standard emergency measures typical of liberal democracies
Intentions of the securitization actor	Not important	Relevant, should be just
Action by the securitization actor	Not important	Relevant, but securitization can be successful without unique solutions, it is important that actions are combined with the securitization movement
Desecuritization	Ends successful securitization – moves issue back to "normal" politics	Should be immediate when the problem moves into the realm of normal politics; for unjust securitizations, it should be automatic

TABLE 4 Comparison of the Securitization Theory and Just Securitization Theory (*cont.*)

	Securitization Theory The Copenhagen School Barry Buzan et al.	**Just Securitization Theory Rita Floyd**
Criteria for successful securitization	Sanctioning the securitization movement through audience acceptance + use of extraordinary measures	Undertaking by the securitization actor of actions combined with (resulting from) the securitization movement
Long-term objectives	Not indicated	Avoid re-securitization of already securitized issues

SOURCE: OWN STUDY

The excessive number of security categories, the securitization of problems that do not pose an existential threat, and the definitional troubles of the concept of security – its voluminousness leads to the question: what security is NOT – threaten to turn it into an empty term that can be easily applied to any issue, thus devoid of any significant meaning (Deudney, 1990b; Freedman, 1998: 53; Walt, 1991). On the other hand, issues securitized decades earlier are (or should be) at least partially transitioning to the desecuritization phase and returning to "normal politics." Others should not be securitized at all, and according to the theory of just securitization, they should be desecuritized as soon as possible.

These doubts arise in various areas of security. Some health security researchers emphasize that placing particular health issues on the security agenda seems largely unrelated to measures of morbidity and mortality. One of the more popular threats – the potential use of pathogens by terrorists (bioterrorism) – accounts for a statistically insignificant percentage of deaths and amplifies doubts about whether such attacks can kill large numbers of people (Fidler and Gostin, 2008: 35; Klotz and Silvester, 2009; Koblenz, 2010: 114–115). Meanwhile, other health issues, such as cigarette smoking, obesity, and infant diarrhea, which cause a substantial number of premature deaths each year, are not securitized (McInnes and Rushton, 2013: 116). McInnes and Rushton, examining the securitization of HIV/AIDS issues, point out that it is not possible to speak in this case of a single process, but rather a collection of separate yet interconnected acts of securitization occurring at multiple levels, involving various securitizing actors and audiences. Some of these "small securitizations" are still ongoing, while others have already concluded. Thus, securitization is not a zero-sum process (as suggested by McDonald, 2008), but rather a continuum, where different actors are situated differently and processes can transpire at varying rates or only partially. The cited authors indicate that in the middle of the first decade of the 21st century, a process of desecuritization of HIV/AIDS issues began at the global level. However, this is not a homogeneous process globally: the disease is only partially securitized outside the United States, meaning that this is not a shift from a clear position, but rather a tendency to move away from framing HIV/AIDS as a security issue – transitioning the matter toward "normal politics" (McInnes and Rushton, 2013: 130).

The securitization of the climate change issue is also questionable. B. Dourley (2016: 16) writes bluntly that the process has failed. "Most actors who talk about climate security call for managing climate change through 'normal' policy decisions. There is a lack of legitimate audiences, the existence of which is necessary to establish a securitization process. There is also little evidence that any emergency measures have been implemented in the climate context." While one can find many examples of securitizers calling for extraordinary measures to address climate change, most call for managing climate change through normal policy channels, such as the UN Framework Convention on Climate Change (UNFCCC) (Oels, 2012: 191). Moreover, the 2008 financial crisis put economic security at the top of the political agenda, and the momentum gained by climate security in the run-up to the 2007 UNFCCC negotiations in Bali was replaced by concerns about the global economy (Oels, 2012: 192) and a focus on it as a potentially more dangerous problem.

Dourley, like Corry (2012), considers the extent to which actions observed in the case of climate change may bear the hallmarks of "riskification," and notes that re-conceptualizing security and distinguishing between threat-based and risk-based security, it is apparent that climate change has fragmented, with current climate change security practices concerned with managing long-standing "conditions of possibility" for determining climate risks (Corry, 2012; Trombetta, 2008; von Lucke, 2015).

Desecuritization does not bypass the issue of societal security. The famous debate on whether minority rights issues should be desecuritized was fought by Paul Roe and Matti Jutila. However, before this occurred, Jef Huysmans (1995), who analyzed strategies of desecuritization in cases where migration transitioned from "extraordinary politics" to "normal politics," noted that some researchers and activists – against the trend of securitization – attempt to describe the migrant and their story in a manner that does not conform to the "security drama." The migrant is portrayed as a person with multiple identities – woman/man, mother/father, teacher/doctor/farmer, etc. – which allows for the deconstruction of migrants' collective identities. Through this process, they cease to be "other" and no longer pose a threat to the "natives." Analyzing this approach, Huysmans concluded that the issue of migration not only needs to be, but above all can be effectively desecuritized (Huysmans, 2000). With numerous migration crises all around the world, it is difficult to find consensus on this issue. Roe (2004) points out that issues of minorities or migration are always part of societal security because they involve key issues of identity and cannot be effectively desecuritized. Jutila (2006), on the other hand, argues that desecuritization is possible as long as there is a prior reconstruction of identity (both of minority groups and the society of the receiving country), leading to multiculturalism, which will allow the issue of minority rights to be transferred from the security to the political (human rights) sphere.

There is also discussion about the possibilities and conditions for the desecuritization of energy, whether in the Middle East (Christou and Adamides, 2013) or in general (through its transition to the problem of "riskification"; Judge and Maltby, 2017). Much has also been written about removing the "stigma" of security from the issue of social security (the argument is made here that in the 21st century there is no need to make the extraordinary that people should be provided with the basics of existence and the chance to develop). However, it should be borne in mind that even the right calls for desecuritization do not automatically lead to its implementation, and that resecuritization of already desecuritized issues may occur, even if it is not "morally justified" (Floyd, 2019).

3 From Threat to Risk: Riskification as Securitization 2.0

The critical voices and approaches presented above are not the only ones that securitization theory has encountered. In the second decade of the 21st century, thirty years after the first publications of the Copenhagen School, there is an emerging observation among authors addressing security issues: it is not threats but risk that is becoming the basis for securitization efforts. They argue that risk is a new security, and they point out that following the trend to securitize various aspects of life comes its second phase – securitization 2.0: the era of identifying risks, detecting potential threats before they occur. This approach is based on the observation that security no longer focuses solely on existential threats, but also on potential threats or risks (van Munster, 2005). The growing importance of risk leads to the extension of securitization across almost all areas of social life, resulting in the implementation of exceptional security measures permanently (rather than temporarily – until the threat is combated) to address potential, hypothetical, and not yet existing threats (risks). Such a transformation in the political logic of security can be observed in both the sphere of action and political discourse (speech acts) (Corry, 2012: 235).

This securitization 2.0, called riskification and described by Olaf Corry in 2012, is based on a focus on "risks" rather than "threats." According to the author, risk policy does not exemplify classic securitization. Securitization, which appeals to existential risks, emphasizes the immediate damage caused by threats, while security appeals based on risk policy are oriented toward the conditions of potential damage and promote long-term preventive management. In this way, they incorporate issues such as citizen health, climate change, and ensuring access to potable water into security concerns. Since potential risks can be identified in every area of human activity (both individual and organized), their functioning can also be considered and described from a riskification perspective (Corry, 2012: 235). This analysis is possible even if the probability of a risk turning into a threat is low. It is sufficient that the action is, as in the case of securitization, approved by society, or by a sufficient part of it, as appropriate to provide security, increasingly understood as protection against risk.

Long-term security assurance projects such as the "war on terror," "preventive warfare," and risk management techniques like registration, control, and profiling are often cited as supporting this thesis (Corry, 2012: 236). The measures implemented by governments and international organizations in response to the COVID-19 pandemic appear to represent another such case.

Proponents of developing "riskification" theories were not the first to point out that risk is an important factor shaping opinion and political decisions. At the end of the 20th century, Ulrich Beck described a "risk society" that first generates and then constantly reflects on catastrophic scenarios, such as nuclear war or climate change. According to the prominent German sociologist, risk is a management technique that refers to the need to secure populations or resources by controlling risk through tools and actions such as surveillance, detention, monitoring, and profiling. Viewing the world through a risk perspective influences strategies for providing security, especially in Western countries. The language describing the current state and development prospects is also changing: it reflects more fear and threats than concerns and recommendations, and more prohibitions than appeals to social responsibility and the common good, Beck concludes (Beck, 2012: 372). Corry elaborates on this thought, pointing out that risk acts as a kind of amplifier of securitization and "activates every technology imaginable in the face of uncertainty," from airport security procedures to full-blown international warfare, such as the 2003 Iraq War (Corry, 2012: 239).

Beck's risk society theory was elevated by Corry to a "global risk society," preoccupied with decentralized and "incalculable" risks that have the potential for catastrophic and irreversible damage. (…) In this global risk society, "the very ideas of control, certainty or security – so important to the first modernity – are collapsing." Risk management, as a way to reduce uncertainty and minimize risk, for example, through insurance, becomes crucial as a "controlled extension of rational action." This approach shifts "security" issues away from deterrence and defense policies toward uncertainty management. War and security are evolving into exercises in risk management, with organizations such as NATO transitioning from territorial security providers to global risk managers (Corry, 2012: 242).

According to other risk theorists inspired by French thinkers such as Foucault and Bourdieu, risk is "a particular rationality of government that acts according to the law of power technology" (Bigo, 2002, 2006). Treating security as a social domain rather than merely a form of discourse, they focus on security practices such as migration procedures and asylum or the war on terror, as the "logic of risk" facilitates the implementation of securitization processes in modern societies, placing new areas of human life under governmental control. Risk is a tool used deliberately by some actors "as a way of governing, to play with anxiety or to stir it up, if it doesn't already exist, to assert their role as providers of protection and security" (Bigo, 2002, 2006; Huysmans, 2006). Society becomes excessively controlled by widespread security practices that use "risk" as a major legitimization tool and strategy.

Thus, the Riskifikation Theory enables the use of security to justify action or inaction in an unlimited way, as long as it is at least partially accepted by society. A real existential threat, imminent and locatable in time, is no longer necessary. It suffices for potential risks, rather than actual threats, to dominate the imagination in the realm of security. As M. Vedby Rasmussen pointed out, there is no perfect security because endless risk scenarios project an uncertain future. Risk, by its very nature, cannot be eradicated or defeated, only managed; thus, an extraordinary policy enacted temporarily in response to real threats, in the case of risk, transforms into a permanent and long-term one. For Vedby Rasmussen, "the goal of security policy is no longer to stop threats, but to 'filter out' unlikely risks from truly dangerous ones." This implies that security measures for risks will be permanent (Vedby Rasmussen, 2006).

The change described can be seen at the institutional level. NATO's transformation from an alliance to a security institution after the collapse of the bipolar world (Haftendorn et al., 1999) results in a qualitative change: the alliance is tasked with countering a common danger, while the institution acts against threats already at the level of risk (Dittgen and Peters, 2001; Navari, 2012: 42; Lake, 2001: 136). This approach has implications for the actions taken by institutions such as NATO. As we have confirmed, risk is something that can be changed and managed, rather than something that can be overcome or eliminated. A threat, on the other hand, cannot be managed but can be destroyed. Hence, while any securitization involves a plan of action to protect a valuable object from a threat (eliminate the threat), riskification implies a plan of action that manages the circumstances of possible damage. The hallmark of securitization is a policy of exceptions, secrecy, extraordinary measures, and time pressure in view of the threat to the security of the protected entity. In the case of riskification, we are dealing with the possibility or threat of damage, which leads to the implementation of special programs of permanent changes that reduce vulnerability and increase the ability to manage potential risks (Corry, 2012: 247–248). A summary of the differences between securitization and riskification is provided in Table 5.

Riskification does not eliminate securitization; the two processes occur in parallel, often intersecting depending on the relationship between threats and risks in a given issue and the level of social anxieties, frustrations, problems, and emotions. Moreover, as critics of riskification point out, it lacks a

TABLE 5 Grammars of security

	Securitization (security politics)	**Riskification (risk politics)**
Grammar	Construction of scenario of direct harm (an existential threat) to a valued referent object.	Construction of conditions of possibility of harm (a risk) to a governance-object.
Political imperative	Plan of action for defense against threat, that is external to referent object	Plan of action to increase governance and resilient of referent object
Performative effects	Legitimation of exceptional measures (secrecy, no-holds-barred action, no trade-offs with security) aiming for survival.	Legitimation of precautionary measures, i.e. inclusion of a safety margin.

SOURCE: CORRY (2012: 249)

transformative dimension but instead reinforces the status quo. For example, developed countries' security agenda focuses heavily on climate change, which is difficult to securitize. However, thanks to Riskification Theory, it can be framed in such a way (by indicating long-term risks) to best protect and promote the interests of these countries (von Lucke et al., 2014: 875). This situation illustrates the subjective nature of the process: "resilience-building" measures are taken only to the extent necessary to manage risks in developed countries, which may not be sufficient to reduce risks in the rest of the world. Riskification can fall into the same trap as securitization, as it primarily serves the interests of wealthy developed countries and further perpetuates the global North-South divide. Additionally, it enables the abuse of emergency precautions more easily than securitisation, especially in situations where the threat is "merely" potentially existential (von Lucke et al., 2014: 872).

CHAPTER 2

Security Discourse Research – Methodological Aspects

Securitization, understood as an intersubjective process of constructing existential threats that, by definition, do not necessarily have to be objective because they are designated as such in the process of speech, must tempt politicians, especially at a time when they craft messages crucial to their electoral success. Ole Wæver's assertion that accuracy is more important in indicating (naming) a threat rather than the truthfulness (objectivity) of this state raises questions about the morality of the whole process. Since whether the threat exists objectively is not important, because the only thing that matters is how it will be presented, by whom, and employing what arguments (Wæver, 1993: 23), it becomes easier not to have scruples about the instrumental inclusion of further issues in the area of security.

There are several perspectives on securitization regarding political tactics, namely:

1. securitization as a deliberate action to justify the use or introduction of extraordinary measures;
2. securitization as an act of speech, where what is said (announced, named) creates new categories and interpretations, leading to decisions and actions (the significance of verbal communication in engaging with the electorate);
3. securitization as a lack of interest (in the classical sense) in the intentions (sincerity) of securitizing actors;
4. securitization as the seemingly democratic process of inclusion in the sphere of security, referring to the "approving audience" – potential voters legitimizing securitization processes with a ballot card (which is ostensible because it results in an undemocratic introduction of extraordinary solutions).

Securitization does not address the state of security but the process of its construction. Therefore, it does not seek a new understanding, nor does it describe the actual state (which, according to the creators of the theory, is not possible in an objective sense); instead, it focuses on the intersubjective process of incorporating and excluding selected issues into and out of the sphere of security. It verifies neither the accuracy of the risks (their objective nature) nor the motives for the securitization of issues. It aims to indicate the conditions of desecuritization, which is challenging in practice and often occurs only partially (when some of the conditions are met) or in selected locations (as seen in the theory of regional conflicts). However, this is a fact that can be observed and analyzed, and it frequently appears in political discourse due to its discursive nature (the act of speech being one of the key stages of the process).

1 Political Discourse and Its Securitization

The above discussion clearly shows that at the center of the securitization process is the speech act, which is a discursive representation of the problem as an existential threat (Fijałkowski, 2013: 97). Unlike threat perception (which is based on the perception of something external), the speech act refers exclusively to itself, to the act of uttering "security." A threat becomes a threat only because it is invoked as such through the utterance of "I threaten" or "I am threatened" (Huysmans, 2006: 24). Thus, as Ole Wæver (1995: 55) stated, "by uttering words something is done. By calling something a security problem, that something becomes one." Hence, securitization is "the intersubjective construction of an existential threat of sufficient importance to have a significant political effect" (Wæver, 2004: 8).

Speech act theory derives from John L. Austin, who distinguished three types of speech acts: locutionary (the creation and articulation of utterances), illocutionary (involving the intentionality of utterances, such as promises and requests), and perlocutionary (which additionally causes some effect on the recipient, for example, by prompting some action) (Austin, 1993). The theory of securitization has generally been based on the illocutionary logic of the speech act. However, Vuori states that for the act of securitization to be considered complete, it must affect the recipient (perlocution). Balzacq also recognizes that full securitization without perlocutionary effects is impossible, since "perlocution is crucial (…) for understanding how a particular public issue can turn into a security problem" (Vuori, 2008: 66; Balzacq, 2011: 4–6).

The securitization movement features a specific rhetorical structure: claim (claim) – warning (warning) – demand (demand), which is supported by evidence backing the claim or warning. A general claim, that something is dangerous (i.e., poses a potentially existential threat), leads to a warning: if no action is taken, the threat will materialise; consequently, there arises a demand: something must be done (Vuori, 2008). The recipient is expected to either support the demand by endorsing the use of extraordinary measures by decision-makers or to take action themselves (e.g., influence decision-makers, refrain from certain actions, or adopt new habits).

The power of the speech act in the context of security is immense. As previously stated, according to the constructivist approach, security is a social construct whose content varies according to different contexts. The absence of arbitrary definitions of security within this approach means that the answer to the question of who recognizes the problems as security issues must be rooted in the rivalries and negotiations of political actors. Participants in these discussions include both elites and societies. Individuals must accept the security narrative presented to them by their political elites. Hence, to create dominant security discourses, political elites compete with one another using "significant ideological inputs" (Weldes et al., 1999: 6), as well as media and pop culture (Croft, 2006).

These processes should be combined with the responsibility of those constructing the discourse. Security policy is not a showy debate in the political arena, but an answer to the question of who can securitize what and under what conditions, as Wæver (1997) reminds us. If mishandled, securitisation can have far-reaching consequences. As long as the political imagination of subjects leans toward viewing the world as a dangerous place where enemies lurk, any situation can be perceived as the result of a conspiracy, which fosters a rhetoric of paranoia. Meanwhile, other dangers continue to be regarded as fates to be submitted to and accepted without discussion (Bigo, 2012: 123). The emergence of new fears and catastrophic scenarios across various aspects of life, such as health (mad cow disease, bird flu, SARS, COVID-19) and communication technologies (the millennium bug, 5G technology), ignites a competition in persuasion. Crafting a specific discourse is a crucial element in this persuasion.

From the perspective of Securitization Theory, based on an effective "speech act," discourse (especially political discourse) emerges as a key element in the process of securitizing issues. It is akin to riskification: long-term risk management demands specific communication, consistently amplifying the seriousness of potential threats, executed to gain favor with the message recipients. Successful securitization is audience-centered, context-dependent, and "is loaded with power" (Balzacq, 2005).

The polymorphous concept of discourse is defined in various ways, depending on the approach used (linguistic, sociological, philosophical, political, and even colloquial; Kujawa, 2009: 44). Discourse is one of those concepts that owes its "cultural ubiquity to semantic openness" (Plumpe, 1988: 330), leading to "semantic underdetermination" (Gajda, 2005: 11). Hence, there remains a keen reference in scholarly works to the etymological meaning of the term (discurrere/discurusus: to move, to run back and forth, as a contribution to the interpretation about the "divergence of rationale," and thus conversation, speech, or address) (Kujawa, 2009).

The study of discourse is linked to the belief that "reality is constructed in communicative processes," and that the world is produced through communicative acts (Gajda, 2010: 253; Grzymała-Kazłowska, 2004: 13; Jabłońska, 2006: 57). Defining discourse as a conversation conducted under certain conditions (Habermas will point out, for example, that the discussion should lead to agreement and be based on the assumption of rationality (Habermas, 2002)), and recognising its various dimensions and consequences (such as the manifestations of power indicated by Foucault (Foucault, 2006; Szacki, 2002: 905) or the domination/violence described by Bourdieu (2005)), it should be noted that discourse encompasses not only language use but also context (Trutkowski, 2004).

As previously noted, the definition and study of discourse depend on the chosen approach. A linguistic discourse analysis will focus on the written text and its structure, a sociological approach will emphasise explanation and description, and a critical approach will aim at changing social reality by solving problems through engaged research practices, such as identifying abuses (Czyżewski et al., 1997: 16–20). Mixed approaches, such as sociolinguistics, are also possible. Consequently, in the literature, one can find definitions stating that discourse is "linguistic and textual productions realized in the form of monologue and dialogue, occurring in spoken and written form, interrelated and interdependent" (Strauss et al., 1989: 603), "a linguistic exchange between sender and receiver, the form of which is determined by social factors" (Leech and Short, 1981), or indicating that discourse is nothing more than "the degree of organization of language into an utterance or the whole process of communication" (Grzegorczykowa, 2008: 40).

The aforementioned J. Habermas pointed out that discourse, considered a social form of communication, is treated primarily as a type of argumentation that must meet certain conditions: the achievement of agreement between the partners of social communication in contentious activities, the exchange of arguments and positions, and the presentation of differing points of view (1986). M. Foucault (2005) sees discourse through the prism of power, as it allows for the determination of subjects' thinking and the "order of things." For him, discourse represents the relationship of knowledge and power and their interaction (Foucault, 2000: 31). P. Ricoeur (1989: 24), on the other hand, points out that discourse is an individualized process of speaking (utterance), usually occurring in a specific temporal order, which includes reference to the external conditions in which it takes place and is always meaningful.

For this book, the definition of discourse proposed by M. Czyżewski (1997: 10) has been adopted. Czyżewski defines it more broadly than strictly linguistic as: "the totality of messages in the social circulation of a colloquial or institutional nature, which consists of verbal (including speech and written texts) and non-verbal communication." This author distinguishes the concept of political discourse from public discourse and the discourse of politics, while pointing out that these areas partially overlap.

He defines public discourse as all publicly available messages, including institutional discourses, discourses associated with specific societies, and mass media discourse (which includes messages disseminated by the media, as well as those arranged for the media). The discourse of politics is that segment of public discourse that encompasses statements made by politicians in their assigned roles within political institutions. This area includes statements made by members of the power elite concerning their political roles and functions. It can occur through the media (Czyżewski et al., 1997: 11–17).

Czyżewski derives his definition of political discourse from T. van Dijk's concept of symbolic elites. These elites include publicists, journalists, writers, scientists, authors of school textbooks, experts, clergy, intellectuals, businesspeople, and politicians. He views these groups and individuals as having direct control over "publicly available knowledge, legitimate beliefs, the shape and content of public discourse." Elites establish the hierarchy of issues or values. They should not be confused with a power elite, which consists of people who hold key positions in the political system, typically expressed in terms of formal authority. In the case of symbolic elites, we rather speak of control over discourse. Czyżewski considers the statements of politicians outside the context of exercising political functions, as well as the statements of all other elite participants on political issues, to be part of political discourse.

It is worth remembering, however, that the distinction between political discourse and the discourse of politics is not unambiguously sharp in the literature. Grinth (1996: 66) defines political discourse as a collection of texts, in spoken and written form, from a well-defined area of political communication. They usually feature a common theme that shapes them and a specific slice of space-time that frames the discourse. Graber points out that we deal with political discourse when "political actors, both in and out of government, communicate on political matters for political purposes" (1981: 196). A broader definition of political discourse can even encompass any discourse or part of a discourse that happens to be about a political topic, such as an informal conversation between friends about an upcoming election or a politicized issue such as abortion (Ädel, 2010: 592). Van Dijk himself, from whose concept of symbolic elites Czyżewski derived his typology, defines political discourse as simply the discourse of politicians, which consists of various statements and actions (2003). Thus, in this sense, political discourse also includes the discourse of politics distinguished by Czyżewski.

2 Organization of Research into Political Discourse

The organization of research into political discourse regarding its securitization required certain assumptions. The first pertains to the selection of material. As previously mentioned, official documents (political party programs and presidential candidates' platforms), speeches (the exposés of Prime Ministers and addresses by Presidents before the National Assembly), and debates available in writing were analyzed. This approach may raise doubts about whether it captures the full dimension of the phenomenon under study. After all, analyzing spontaneous statements may reveal the true intentions of securitization actors that they might not wish to disclose to voters. This is not an area of research that should be disregarded, but the assumption of this book was based on the belief that it is crucial to examine what individual parties and their candidates say in the area of security, rather than only when prompted about it, and how significant these issues are compared to other topics raised during the campaign. It is assumed that securitization is a conscious process whereby selected problems are presented as existential risks that cannot be solved through ordinary means. Such a process necessitates reflection and preparation; thus, the analysis of electoral programs, exposés, or parliamentary speeches during planned rather than ad hoc debates is essential. The manner in which a given issue is addressed in such texts or speeches (e.g., the migration crisis presented as a threat to the security of Poles) is not coincidental; the subject matter and its form are not the result of linguistic error or a slip of the tongue, etc. These deliberate actions by political leaders and parties are crucial in assessing the securitization of Polish political discourse.

This approach is grounded in the experiences of other researchers involved in analyzing the electoral programs of political parties. As Cześnik notes (2011: 7), "the advantage of analyzing the content of party documents is that it refers directly to the official position of the party (…) because electoral programs are developed during intra-party discussions and are often the result of compromise between party factions." They contain the content that the party, as a collective actor in the political process, wants to communicate to voters. Their position, placement in the program, and the amount of space devoted usually reflect the importance of a given subject for a party or a political group

(Robertson, 1976). Party programs are the only documents that can be considered holistic visions of the social order developed by political parties – the only political institutions that possess the potential to implement them if they win an election (Budge, 1994). Their importance is enhanced by the fact that they are approved by party decision-making bodies to be executed in the event of attaining power (Budge et al., 2001; Budge et al., 2006). As Woźniak (2017: 41) states, these are studies that place the party within a concrete, historical, and ideological context, and thus, they should be taken seriously in scientific discourse as descriptions of the desired future of the nation, state, and society. Therefore, although this approach is not without flaws, it seems appropriate to examine the intentions of politicians and how they communicate and bolster it.

The study of political programs and other official party documents has a long tradition. At least since the 1970s, political scientists have used both quantitative and qualitative content analysis as a technique that leads to discovering crucial threads communicated to voters by political broadcasters. In quantitative research supported by statistical programs, reference is made to the frequency of specific categories, the formal structure of the message, keywords, and the amount of space devoted to the selected subject. These elements enable the construction of an image of discourse; however, it should be remembered that content related to security does not always focus on this concept. It can be found in the excerpts of the analyzed texts, where the key word is sovereignty or independence, justice or equality, rather than "more obvious" threats, fear, and danger. Therefore, only the qualitative analysis of the presented content, its proper qualification and systematization, and an approach broader than a percentage one (for example, indicating the place of a given issue in the program – the order of appearance) will provide a full picture of the analyzed phenomenon. Finally, it should be borne in mind that in times of online campaigns and social media, the nature and presentation of the electoral program are changing. From long documents that carefully analyze selected issues, it evolves toward keyword leaflets and posters or banners that are easy to present on the Internet. Obviously, some parties continue to issue traditional program documents; however, sometimes the party's or candidate's program does not take a structured form at all. It also happens that the party or candidate does not present program assumptions in any compact form except for posters/banners formatted under the requirements of social media. Excluding them from the analysis of the program solely because of form would be a significant negligence.

For years, political scientists have been seeking solutions for a systematic analysis of the content of electoral programs that would avoid the allegation of superficiality (Ginsberg, 1976). One of the most interesting concepts was the proposal of the 1979 Manifesto Research Group (MRG) (the name evolved in 1989 into the Comparative Manifesto Project (CMP); currently: Manifesto Research on Political Representation, MARPOR 2020), which operates to this day and addresses this issue as part of long-term research grants, involving Ian Budge, David Robertson, Peter Mair, and Michael Laver. They proposed an innovative method for coding content in political programs, which facilitates further quantitative analysis. This approach divides content into quasi-sentences, each assigned a single code that primarily represents the content (the number of codes changes as the code table is updated). Although it remains the most popular method of content analysis in political sciences (Volkens et al., 2009), it faces substantial criticism, mainly for neglecting context, interpreting language literally (thus overlooking nuances and subtler forms of expression), or for its asymmetrical coding approach (Markowski, 2002; Cześnik, 2011: 14–15).

Regarding the usefulness of the methodology for researching the securitization of policy programs, it is limited. It allows examination of how much space in the party's programs is devoted to specific security-related categories or problems, but it cannot indicate whether this relates to the securitization or merely the politicization of these issues. The method assesses the sheer act of discussing a problem or attitude in a positive or negative context, but does not consider other aspects of discourse. The answer to whether a securitization has occurred can only be provided through a critical analysis of the text (discourse).

Quantitative research aims to show the extent of space dedicated to security categories in the electoral programs of Polish parties and presidential candidates. In terms of quantity, the exposés of the Prime Ministers (from Tadeusz Mazowiecki in 1989 to the third exposé of Donald Tusk in 2023) and speeches given by the Presidents before the National Assembly during their inauguration and assumption of control over the Polish Armed Forces (from Wojciech Jaruzelski in 1989 to the second swearing-in ceremony of Andrzej Duda in 2020) will be analyzed. Quantitative research will address how much attention is given in Polish political discourse to security issues and whether changes can be observed in this area over the past 30 years. However, it will not provide answers to whether the process of forming (and possibly expanding) security categories involves securitization or politicalization (no proposals for extraordinary solutions, no need to accept voters).

Therefore, the second part of the research involved a qualitative, critical analysis of discourse, which helped identify evidence of the securitization process in the analyzed speeches and texts. They will be examined in both the programs and the exposés of the Prime Ministers, as well as the speeches of the Presidents sworn into office and the five parliamentary debates selected for analysis:

1. 26th March 2003. *Information of the Government on the participation of the Polish military contingent in the composition of the forces of the international coalition to enforce Iraq's compliance with UN Security Council resolutions* (with discussion);
2. 24th May 2012. *Information of the Minister of National Defense on the strategy for withdrawal Polish troops of Afghanistan* (with discussion);
3. 16 September 2015. *Information of the Prime Minister on the migration crisis in Europe and its repercussions for Poland* (with discussion);
4. 2nd March 2020. *Information of the Government on actions taken and the state of the country's preparations related to epidemic threats in Poland and Europe* (with discussion);
5. year 2022. 11 discussions and debates related to the presentation of *The bill on assistance to citizens of Ukraine in connection with the armed conflict on the territory of this country and its changes* (with discussion) (08.03.2022; 09.03.2022; 23.03.2022; 06.04.2022; 07.04.2022; 27.04.2022; 11.05.2022; 26.05.2022; 08.06.2022; 13.12.2022; 14.12.2022).

Only such an analysis enables an attempt to address the question regarding the process of securitization and its scale in Polish political discourse. Qualitative discourse analysis, as noted by Wodak (2001, 2008, 2011), Meyer (2001), and Krippendorff (2004), facilitates an interdisciplinary approach to the content under examination. Following van Dijk (1990: 164), this approach places the text in context to understand its meaning within specific circumstances. It is essential to highlight this method, as there is a distinction between Anglo-American and European scholarship: first, Anglo-American researchers often see discourse synonymously with text, while French researchers are more focused on the connection between language and thought (Wodak and Krzyżanowski, 2017: 5). Following Wodak and Krzyżanowski (2017) and Lemke (1995: 7), I distinguish between discourse and text in this volume and understand it as a different, more abstract level, with text being regarded as a longer piece of writing.

The collection of political programmes and transcriptions of political speeches and debates selected for analysis was coded to reflect the content in the context of security issues raised by the authors. Coding was based on qualitative analysis of the texts: the selection of fragments marked as relating to security issues was not limited to the occurrence of the word "security," as this would not only be insufficient for studying the securitization of Polish political discourse but could also lead to false conclusions. The text analysis revealed that some fragments relate to security issues even though they do not contain the word "security." Moreover, there were texts in which the use of the word "security" was not enough to classify them as related to security issues.

Next, selected fragments of the texts were coded according to their affiliation with the discourse on security sectors, based on the classification presented in Table 1, as well as sub-sectors or thematic areas within the indicated sectors.

3 Quantitative Analysis of Speeches of Polish Presidents and Prime Ministers

Prime ministers' exposés and presidents' speeches before the National Assembly serve as valuable comparative material for studying trends in political discourse, particularly concerning security issues. Each prime minister who presents his government and its main ideas chooses, with great care, the topics, the context in which he presents the issues, and their order. These speeches are prepared and well thought out, so it is legitimate to assume that any securitization treatments included in them are intentional. The same holds true for the presidents' speeches at their swearing-in ceremonies: at this point, each outlines the main lines of his future policies, areas of interest, and announces key actions. As with prime ministers, spontaneous, unintentional plots are hard to come by.

Although presidential speeches are about half as long as those of prime ministers, the portion of content dedicated to security issues is similar. For prime ministers, this percentage varies from 5.43 (W. Pawlak's exposé in 1992) to 34.04 for W. Cimoszewicz. Among presidents, L. Walesa allocated the least to security concerns in 1990 (6.17 percent), while B. Komorowski in 2010 allocated 27.72 percent. In the case of prime ministers, the percentage line resembles a sine wave, whereas for presidents, there is a noticeable increase and stabilization around 25 percent.

Comparisons between left- and right-wing prime ministers and presidents do not indicate a single trend. In the case of prime ministers, government leaders from right and center-right backgrounds, on average, devoted 15.9 percent of their exposés to security, while representatives of the left devoted 22.26 percent. However, for presidents, the situation is exactly the opposite: in the speeches of heads of state from the right, security issues occupied an average of 20.45 percent of the programs (the statistics were significantly underestimated by L. Walesa speech), while those from left-wing backgrounds accounted for 9.96 percent. Figures 3–6 present a summary of the numbers and percentages of the analyzed speeches.

Table 6 and Figure 7 present a chronological comparison of exposés and speeches by presidents to determine whether there is convergence in the amount of space allocated to selected issues based on historical context. Four

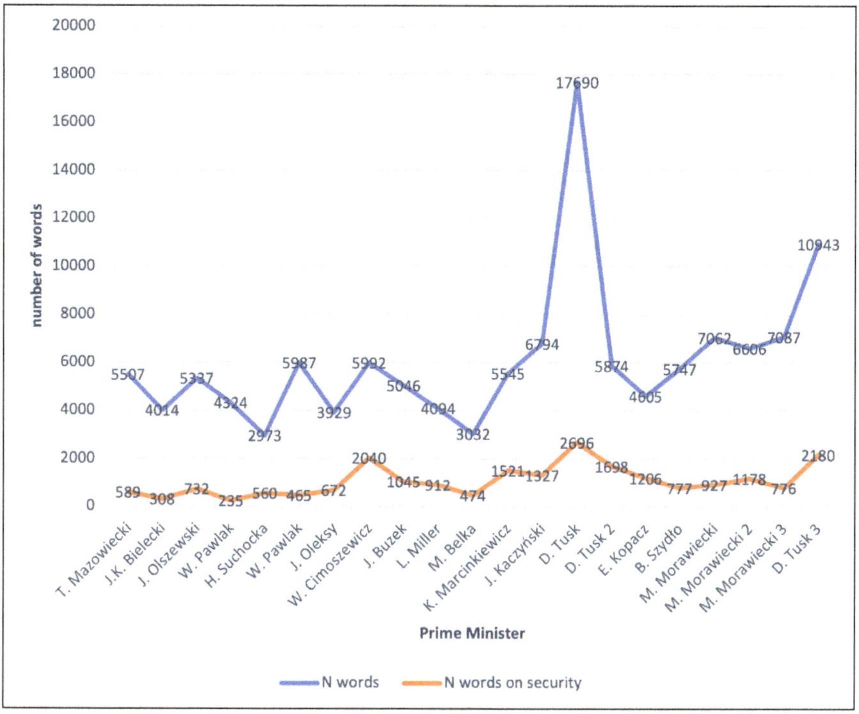

FIGURE 3 Statistical summary of prime ministers' exposés
SOURCE: OWN STUDY

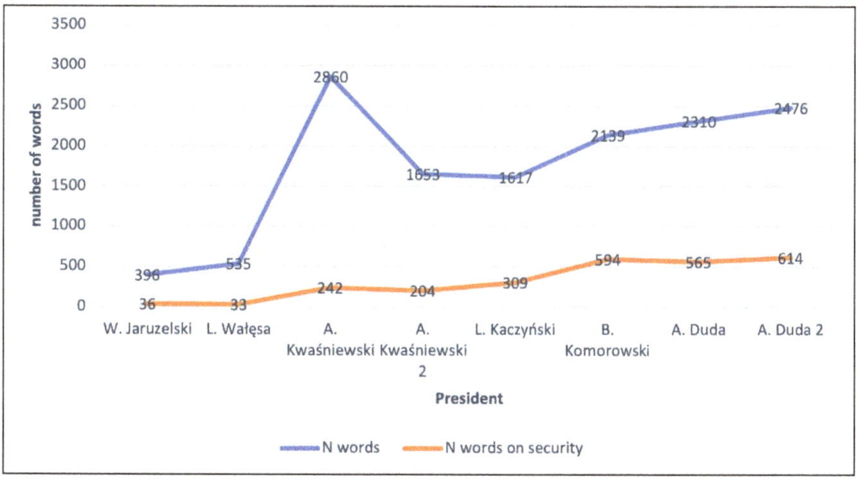

FIGURE 4 Statistical summary of presidents' speeches before the National Assembly
SOURCE: OWN STUDY

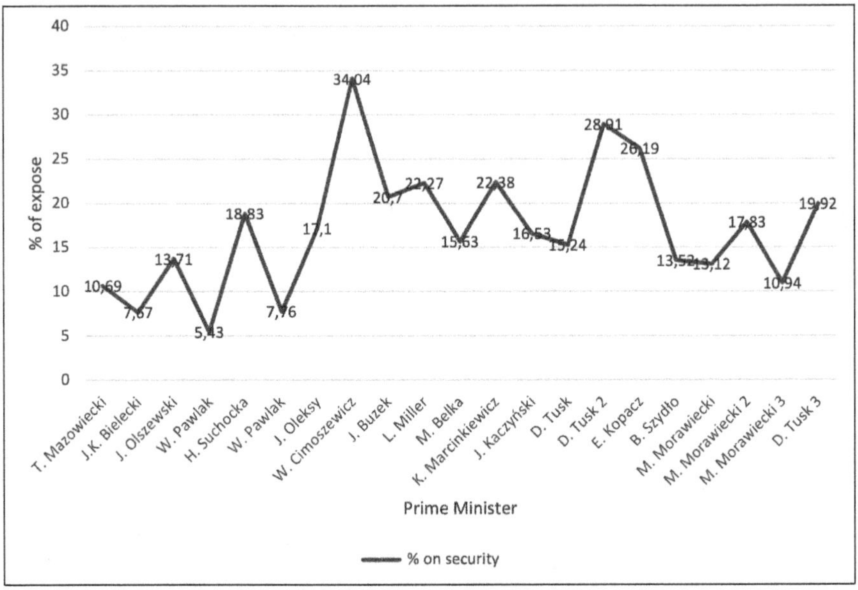

FIGURE 5 Percentage of speeches by prime ministers on security issues
SOURCE: OWN STUDY

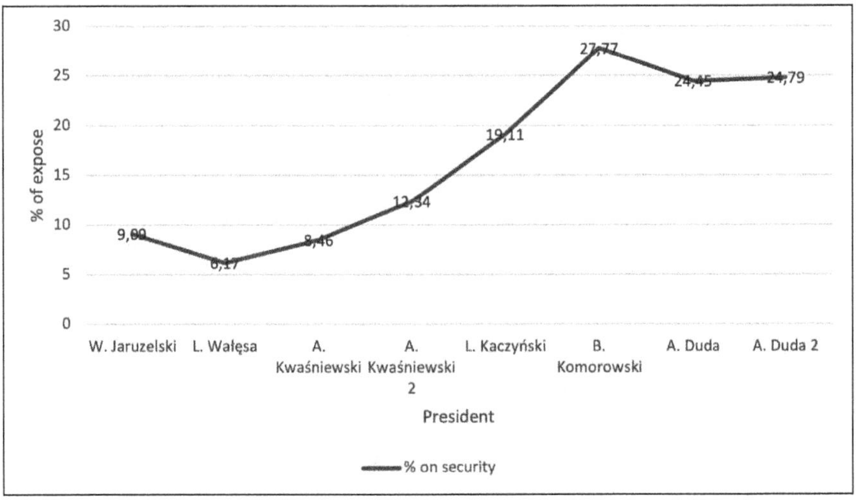

FIGURE 6 Percentage of speeches by presidents on security issues
SOURCE: OWN STUDY

TABLE 6 Comparative percentage of space devoted to security issues in the speeches of presidents and prime ministers in the years 1989–2023

Year	President		Prime Minister	
1989	**9,09**	**W. Jaruzelski**	**10,69**	**T. Mazowiecki**
1990	6,17	L. Wałęsa		
1991			**7,67**	**J. K. Bielecki**
			13,71	J. Olszewski
1992			5,43	W. Pawlak
			18,83	H. Suchocka
1993			7,76	W. Pawlak
1994				
1995	8,46	A. Kwaśniewski	17,1	J. Oleksy
1996			34,04	W. Cimoszewicz
1997			20,70	J. Buzek
1998				
1999				
2000	12,34	A. Kwaśniewski		
2001			22,27	L. Miller
2002				
2003				
2004			15,63	M. Belka
2005	**19,11**	**L. Kaczyński**	**22,38**	**K. Marcinkiewicz**
2006			16,53	J. Kaczyński
2007			15,24	D. Tusk
2008				
2009				
2010	27,77	**B. Komorowski**		
2011			**28,91**	**D. Tusk**
2012				
2013				
2014			26,19	E. Kopacz
2015	**24,45**	**A. Duda**	13,52	B. Szydło
2016				
2017			13,12	M. Morawiecki
2018				
2019			17,83	M. Morawiecki

TABLE 6 Comparative percentage of space devoted to security issues in the speeches (*cont.*)

Year	President		Prime Minister	
2020	24,79	A. Duda		
2021				
2022				
2023			10,94	M. Morawiecki
			19,92	D. Tusk

SOURCE: OWN STUDY

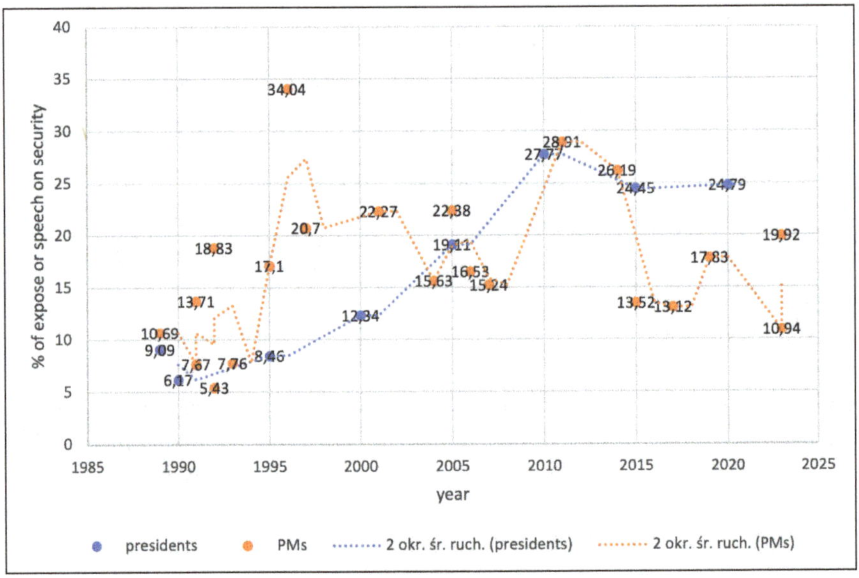

FIGURE 7 Points of convergence in the amount of space devoted to security issues in the speeches of presidents and prime ministers
SOURCE: OWN STUDY

moments of convergence (1989–1991; 2005, 2010, 2014–2015) and three periods of divergence (1991–2004, 2006–2009, 2016–2023) are evident. In 2010 and 2014–2015, a notable convergence occurred, with both exposés and presidential speeches dedicating one-quarter of their space (or slightly more) to security issues. In the periods of 1989–1991, this convergence fluctuated around 10 percent, while in 2005, it was approximately 20 percent.

Another important aspect is the security categories addressed in the speeches of prime ministers and presidents, summarized in Figures 8 and 9. Each

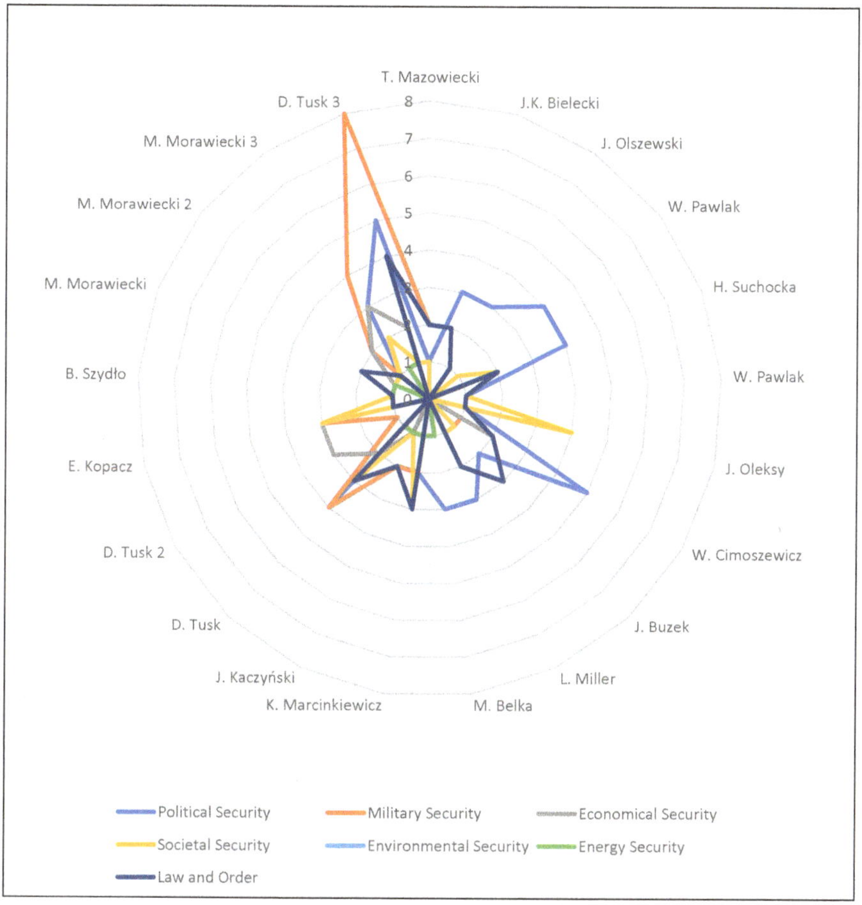

FIGURE 8 Distribution of the frequency of occurrence of various security categories in prime ministers' exposés from 1989 to 2023
SOURCE: OWN STUDY

reference to a category count as one point. If more than one issue from a given category is raised, it is coded as another thread and marked as an additional point on the chart.

By juxtaposing the two charts ("prime ministerial" and "presidential"), it becomes evident that prime ministers raised issues from a broader range of categories than presidents, who made no references to environmental or economic (financial) security at all. In contrast, the dominant category addressed in the speeches for both groups is political security with respect to international relations. The second category is military security, and the third is law and order. Thus, this indicates a correspondence to the prioritization of security categories (see Figure 10).

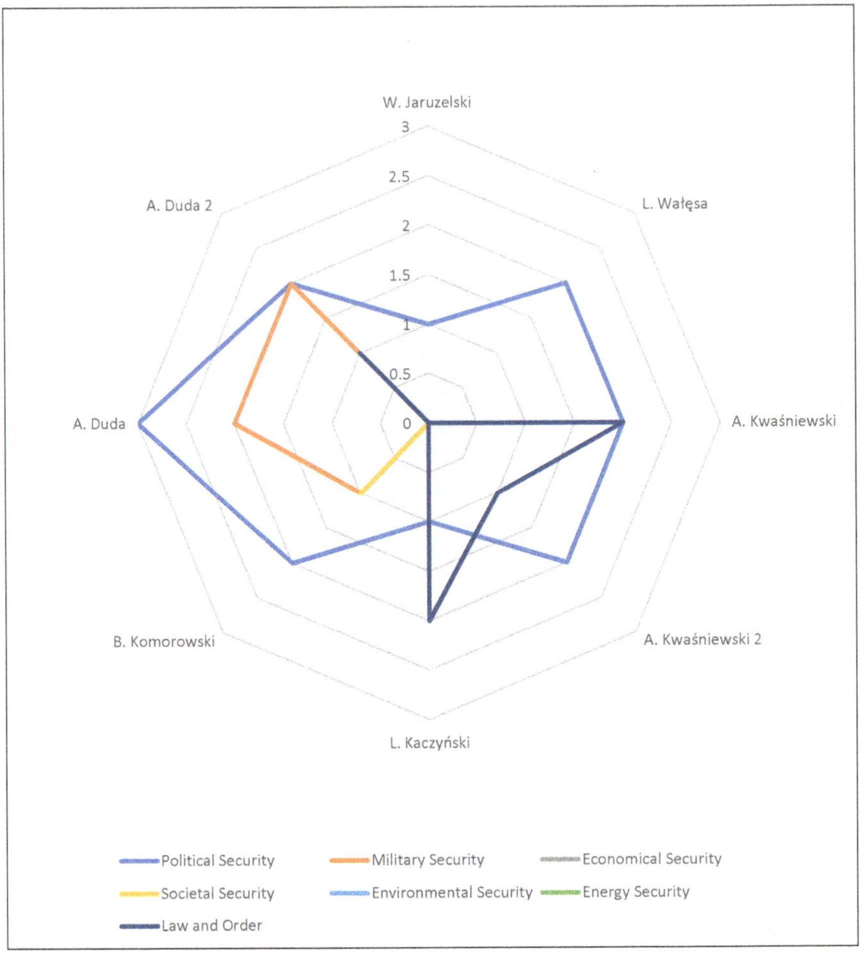

FIGURE 9 Distribution of the frequency of occurrence of various security categories in presidents' exposés occurrences from 1989 to 2020
SOURCE: OWN STUDY

Table 7 presents summaries of the categories addressed in both groups.

Politicians on the left or with leftist roots who hold the position of prime minister more frequently than right-wing politicians refer to political security in its international aspect (relations with other states, membership in organizations). However, in the case of presidents, the situation is reversed. They do not exhibit a greater tendency to address social security issues compared to their right-wing counterparts, which is traditionally seen as a "specialization" of the left. On the contrary, it is the right-wing prime ministers and presidents who more often tackle these issues. Right-wing politicians in both roles also raise military security concerns significantly more than the left.

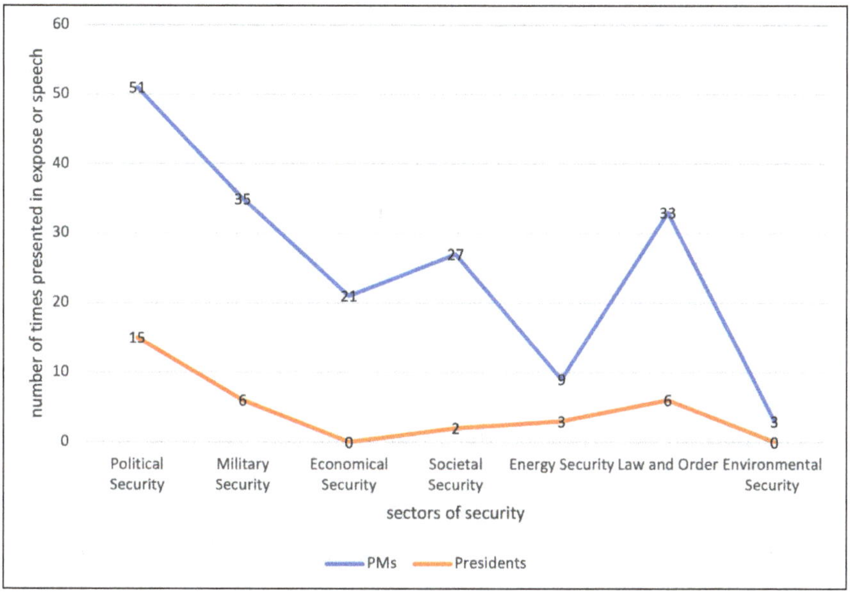

FIGURE 10 Distribution of the frequency of occurrence of various security categories in the speeches of presidents and prime ministers from 1989 to 2023
SOURCE: OWN STUDY

TABLE 7 Comparison of the occurrence of selected security categories in the speeches of presidents and prime ministers

Security category	Political security	Military security	Economical security	Societal security	Energy security	Law and order	Environmental security
PMs total	51	35	21	27	9	33	3
Left/central-left	12	3	3	4	1	5	0
Right/central-right	39	32	18	23	8	28	3
Presidents total	15	6	0	2	3	6	0
Left/central-left	5	1	0	0	0	3	0
Right/central-right	10	5	0	2	3	3	0

SOURCE: OWN STUDY

CHAPTER 3

Military Security Discourse

Military security has been a concern for Polish political parties and groups since 1989. While its significance has not diminished, the emphasis and prioritization of related content have changed. Nevertheless, three key areas can be distinguished throughout the examined period: the first involves (1) allied issues, such as leaving the Warsaw Pact (with the withdrawal of Soviet troops from Poland) and joining NATO, along with the "quality" of membership in that organization; the second (2) focuses on the transformation of the Polish Armed Forces and their role in a democratic state, particularly their participation in foreign operations; and following the annexation of Crimea by Russia (3), addressing the need to build an army prepared to respond to potential conventional military threats to Poland and military aggression. This chapter analyzes political discourse regarding army development, military spending, and the socially controversial involvement in international military operations.

1 Discourse on Army and Military Spendings

The first stage of army modernization from 1989 to 1991 primarily focused on its depoliticization, the introduction of democratic control over the military, and the replacement of command personnel. The next stage emphasized humanization and reducing the duration of compulsory conscription. The third stage concentrated on improving training quality and enhancing defense capabilities. Most parties participating in the elections agreed on the necessity of these principles, which meant that extraordinary measures and securitization were unnecessary. Although these issues ranked high among priorities, they fell under the rules of ordinary politics. There was also a consensus on the importance of allocating adequate financial resources for this purpose.

Between 1989 and 1992, five prime ministers delivered exposés. They were unanimous about the future of the military. Tadeusz Mazowiecki spoke of "an apolitical and respected army, in which the service must be humanized"; Jan Krzysztof Bielecki spoke of the "nationalization of the armed forces and rebuilding patriotic credibility." Jan Olszewski generally mentioned the necessity to reform MON and rebuild the military. Hanna Suchocka announced the reform of the Ministry of Defense and the reduction of the army to 250,000 while ensuring, "There will be a place in the Polish Army for every valuable

DOI:10.1163/9789004745414_005

soldier." All prime ministers referred much more to the international situation and alliances and focused less on internal military changes.

During the analyzed period, two politicians held the office of president: Wojciech Jaruzelski and Lech Wałęsa. Neither addressed the issue of the military and its transformation in their speeches to the National Assembly. It was only upon assuming supremacy over the Armed Forces that Lech Wałęsa spoke of the need to "return the military to the nation" (Wałęsa, 1990).

The 1993 election campaign didn't differ from the previous one. The greatest emphasis was placed on reforms and their financing. The first demands arose for transforming the Polish army into a professional force in preparation for future NATO enlargement. Prime Minister Włodzimierz Cimoszewicz declared in 1996: "The government will support the modernization of the armed forces, providing for a systematic increase in outlays for research and technology acquisition, as well as for the purchase of armaments and military equipment" (Cimoszewicz, 1996). Upon assuming the office of president in 1995, Aleksander Kwaśniewski stated before the National Assembly that, "as head of the armed forces of the Republic, he would not neglect anything that could realistically serve to modernize the national defense system."

Two years before joining NATO, in the 1997 parliamentary campaign, attention was primarily paid to reforming the armed forces for compatibility with the Alliance armies. As previously mentioned, there were no significant differences between the parties in their opinions on the direction of the proposed changes. Instead, there was a political concern, that membership in NATO is a strategic goal for the next years, so military issues were politicized, not securitized. Jerzy Buzek, who took over the Prime Minister's portfolio after the elections, declared: "We will continue and intensify efforts to adapt the Polish armed forces to the requirements of Poland's participation in NATO. An essential part of the state's defense system will be armed forces of smaller numbers, with greater combat value" (Buzek, 1997). In 1999, Poland became an official member of NATO. Strategic goal was achieved without securitization, thanks to political solidarity. From this moment, any declaration related to Poland's military security will refer to the Alliance as a security guarantee. However, the consciousness that it is necessary to be an active member of NATO was present in the statements of politicians. Inaugurating his second presidential term in 2000, Aleksander Kwaśniewski, meeting with soldiers, noted: "The army's affairs deserve constant, continuous and even greater care than before."

In the 2001 campaign, the electoral programs did not address military issues. They focused on the political aspects of NATO membership and applying for EU membership. Leszek Miller, who took over as Prime Minister, recognized that further modernization of the army was necessary, as these issues transcended political divisions: "Despite the poor situation of public finances, the reconstruction and modernization of the armed forces will continue" (Miller, 2001). The 2005 campaign referred to security issues strongly, but mostly internally. Regarding the military, further reform in line with NATO requirements, technical modernization toward a professional army, and allied cooperation were advocated. Prawo i Sprawiedliwość (PiS, Law and Justice) devoted significant attention to military affairs in its program, addressing selected aspects in detail. It did not present anything unusual: "PiS will support the creation of modern operational forces capable of undertaking the full spectrum of expeditionary missions alongside our allies. Achieving these goals may entail an increase in defense spending" (PiS, 2005). The victorious party's political competitors did not devote much space to military matters in their programs. SLD, PO, and PSL did not address this issue at all; LPR did so in a perfunctory manner: "Poland, as a member of NATO, should have a professional, strong army – mobile and equipped with weapons, ammunition, and military equipment primarily produced in Poland, consistent with NATO standards" (LPR, 2005). The campaign prior to the 2007 elections devoted more emphasis to questions regarding the army's condition and its future. The key words of the period were transformation, modernization, professionalization, and modernization leap. Still, it was an aspect of "normal politics," without securitization. Change was included in all programs addressing the state of the armed forces, but calls for exceptional measures did not accompany it. PiS referred primarily to the reforms that had begun and described the following: "The process of restructuring the Polish armed forces is underway. We are ensuring that Polish soldiers participating in military missions are provided with equipment that best protects their health and lives" (PiS, 2007). Platforma Obywatelska (PO, Civic Platform) proposed similar modernizing solutions. It declared that: "the basic element of the national security system is the Polish Armed Forces. It is necessary to rationally transform them adapting them to new threats and challenges" (PO, 2007). PSL only laconically referred to the universal nature of military service: "We will spread a sense of obligation to perform reformed and rational military service in order to ensure universal participation in protecting the nation's external security" (PSL, 2007).

Taking over the Prime Minister's position in 2007, Donald Tusk touched on every conceivable aspect of security in his unusually long exposé, addressing its military dimension primarily in the context of a thoroughly professional army: "A professional army is needed for Poland as soon as possible, because the nature of threats to our security has changed. All of us, as we sit in this room, want an army that will provide us with the greatest degree of security" (Tusk, 2007). This statement by the prime minister is typical for the first decade of the 21st century – we are in NATO, there is no war near our borders, and the army should be fully professional, with a maximum of 100,000 personnel, rather than being based on universal forced conscription. There is no existential threat, no extraordinary measures, and no securitisation. Even though in the 2011 elections, the programs of the political parties detailed the condition of the army and its future, assessing both the reforms introduced by the PO-PSL coalition and postulating their own, the discussion did not cross the line of normal political debate. Opposition parties criticised the state of the Armed Forces, while coalition parties spoke of further directions for modernisation. PiS called for exceptional spending on the army. Still, they focused more on the NATO requirements than any internal need driven by new existential threat: "In order to ensure defense 'self-reliance,' Poland should increase the defense budget to 2% of GDP. In case of a tightening of the political-military situation near Poland, we must expect a significant increase in defense spending" (PiS, 2011). Other programs, such as the one prepared by SLD, which addressed military security in 3819 words, were comprehensive, but it is difficult to pinpoint the priority in this area – for it treated with equal attentiveness strategic decisions to change the approach to the use of the Armed Forces (from expeditionary to defense of the national territory) as it did the presence of chaplains in the army. The coalition parties (PO and PSL) declared the continuation of the policy they had begun about the army:

> We have built a 100,000-strong professional army and established a National Reserve Force. In place of compulsory basic military service and military training, we introduced voluntary contract professional military service and the possibility for women to serve. (...) We have begun a thorough modernization of the army's equipment. (...) Thanks to this, Poland does not differ in terms of the level of equipment from its allies. (PO, 2011)

Donald Tusk, taking over as Prime Minister for the second time, focused on overcoming the economic crisis. Despite this, he announced that military spending would remain at 2 per cent, as his opponents suggested. His successor, Ewa Kopacz, stood by these announcements, adding a reference to the care of veterans (Kopacz, 2014).

The year 2014 transformed the geopolitical landscape in Poland's neighborhood: Crimea was annexed, and separatist movements were orchestrated in Ukraine's Donbas region. It became clear to political decision-makers that the era of military expeditions beyond Poland's borders had come to a close and that the focus should shift to rebuilding the country's capability to defend its territory. Consequently, politicians began to discuss a special situation and the necessity of strengthening the army; however, widespread agreement on this course of action resulted in the issue largely remaining in the realm of normal politics. Even after 2022, securitization will not be evident due to broad public acceptance of increased military spending and interest in military training for civilians.

Before the 2015 parliamentary campaign began, there was a change in the Presidential Palace. Standing before the National Assembly to take the oath of office, President-elect Andrzej Duda indicated, following the described narrative: Military aspect of security is the most important in the current situation. "The most essential thing in this respect is to build a strong, well-equipped Polish army" (Duda, 2015). Standing in front of the soldiers, during the ceremony of taking command over the armed forces, he added: "Today, the duty of the Polish authorities towards you and the Republic of Poland is to modernize the Polish army, arm it with the finest weapons, so that you can fight more efficiently" (Duda, 2015).

In the 2015 campaign, the opposition devoted more attention to the army and defense spending than the coalition parties. The PiS programme was the most extensive. The party discussed necessary changes in defense doctrine and the use of the Armed Forces in the context of current threats (the annexation of Crimea), emphasizing the need to prepare the military to defend the national territory. This marked a shift from the previous strong support for the participation of the Polish Armed Forces in foreign operations:

> The international situation is changeable, and an army model that makes sense in certain circumstances loses its meaning in others. Poland is experiencing a geopolitical shift in its security environment. (…) The priority of building the expeditionary capabilities of the Polish Army, which was binding in previous decades, must be discarded. (PiS, 2015)

The party highlighted the need to increase the army's funding and size and to gain public acceptance for these measures due to the new threat posed by 'geopolitical change.' PiS acknowledged that the revolutionary change in the army's operations (including the proposal to establish the Territorial Defense Forces) required extraordinary measures (e.g., raising expenditure to 2.5 per cent of GDP and enacting a special law to support the Polish defense industry), which could be seen as a securitizing move. PiS regarded these demands as an urgent task to be implemented immediately and declared that military security would be its priority (occupying the first position in the group's programme). Similarly, the Kukiz15 party addressed the new external threat. It also called for the strengthening of existing operational forces and the establishment of the

Territorial Defense Forces (Kukiz15). Nowoczesna, a new party on the political scene, mentioned the army briefly in the context of maintaining defense spending at 2 percent of GDP and developing the defense sector. Lewica was similarly brief on the defense issue, focusing on the defense industry and advocating for pay raises for soldiers (Zjednoczona Lewica 2015).

PO, which was fighting for re-election, did not duplicate the opposition's alarmist style. In its program, it relied on the continuation of measures implemented since 2007 by the government co-founded by the party. It spoke of modernization, the promotion of pro-defense and patriotic attitudes, and support for veterans. The coalition partner, PSL, presented its program for the military with slogans related to the size of the army, professionalization, and adequate military spending. They did not properly indicate public sentiment regarding the growing threat of military aggression in the region and lost the elections.

The 2019 parliamentary campaign was less focused on military affairs than the previous one. PiS (part of the Zjednoczona Prawica, ZP coalition), this time in the role of the party seeking re-election, did not criticize the Polish Armed Forces but announced the continuation of previously taken measures: "Under conditions of destabilization in the international order, Poland needs an expanded and modernized Armed Forces. We plan to increase defense spending to 2.5% in 2030. In the face of new challenges, we will increase the size of the operational army to at least 150,000 troops" (ZP, 2019). The opposition KO punctuated five years of ZP rule and called for restoring the state of affairs from the time it handed over the government (KO, 2019). Konfederacja laconically mentioned a modern army as a component of state sovereignty. Taking over the Prime Minister's portfolio for the second time, Mateusz Morawiecki also focused on the number of people dying on Polish roads rather than on the state of the armed forces. He reaffirmed only maintaining two percent of GDP spending on the military.

Taking place more than a year after the Russian aggression against Ukraine, the 2023 parliamentary election campaign had to address military issues. However, the intensity of this issue in the programs varied. There were parties or coalitions, such as Trzecia Droga, that focused solely on strengthening Poland's defense capabilities while expanding and modernizing the army: "Let's be strong with what we produce: at least 50 percent of military equipment should be purchased from companies and factories that produce in Poland" (Trzecia Droga, 2023). Similarly, Konfederacja has stated generally that Poland needs efficient armament spending and a larger army, but it views increased military security in its longstanding demand for expanded citizens' access to firearms (Konfederacja, 2023). The program referenced "the ongoing warfare taking place across our eastern border," in the context of which it called for building national defense capabilities when allied arrangements fail. In contrast, Lewica advocated for developing military capabilities in cooperation with the EU and NATO. The army was "to be professional, strong and voluntary – without reinstating conscription." Lewica also pledged to "continue modernizing the army's equipment and adapting it to today's needs," as well as auditing the decisions of its predecessors. The content of the program did not suggest that Poland was facing special times or circumstances. The KO program was similarly structured, referencing allied relations and the continuation of equipment purchases following an audit of contracts made by the ZP government.

Against this backdrop, the PiS program looked as if it had been transported from another dimension. The party enumerated in detail across dozens of pages what equipment it had purchased to strengthen the Polish Armed Forces and what it had done to defend its borders. Above all, it drew a catastrophic vision of a world threatened by war: "After 2014 and the Russian aggression against Crimea and Ukraine, a blatant deterioration of security on the Old Continent began. This dangerous situation was accompanied by a radical weakening of our military capabilities" (PiS, 2023). Among the successes of the two-term government in power was the doubling of the army's size, which included the establishment of the Territorial Defense Forces, technical modernization, and the purchase of new equipment. This resulted in an increase in the state's military spending to 4 percent of GDP annually. Although the tone of this part of the program was dramatic, it is difficult to identify an act of securitization here. This difficulty arises because there are no calls for extraordinary measures; the entire operation is taking place within the framework of normal budget planning and the standard process of financing the armed forces.

The issue of the state, use, and reform of the Armed Forces appeared in all campaigns and most of the speeches by the analysed presidents and prime ministers. It was repeatedly regarded as an important, priority, necessary, and urgent issue. However, it did not transcend the boundaries of ordinary politics: the discussion of modernization, development, and changes in the army occurred within the ordinary political process. This probably happened because, despite differing approaches from various groups, the fundamental necessity of strengthening the army was not questioned. The parties opposing the PiS only spoke about reviewing Zjednoczona Prawica's signed contracts for potential fraud or corruption, but did not assert that the purchases themselves were unnecessary. The army's modernization during Poland's most challenging period after 1989 was more about stability and the state budget's capacity than about military considerations *per se*.

This example shows that discourse on security does not require securitization: issues that are naturally security concerns, such as the condition of the army, even in times of growing external military threats, can be addressed within the framework of "normal politics" as long as there is political and social consensus on the direction of action – there is no need to convince the public of an existential threat and extraordinary measures; they become understandable and part of the normal political process.

2 Discourse on Foreign Military Operations

The topic of the strategy regarding the Polish military's participation in foreign operations was not mentioned in election programs until Poland joined NATO and its involvement in Kosovo, Afghanistan, and Iraq began. This was, of course, due to the fact that until 1999, Poland was engaged in UN operations, which were of a different nature than those of NATO and involved fewer soldier casualties. Only the experience in Iraq prompted the parties to consider this area.

Marek Belka was the first prime minister to address the issue in his exposé in 2004. However, these were merely brief comments on remaining in Iraq as an allied commitment. A year later, Kazimierz Marcinkiewicz presented the realization of these commitments in a slightly broader context: "We will use the experience and people who passed the test in Iraq to create an expeditionary brigade capable of operating abroad" (Marcinkiewicz, 2005). In 2007, the PO wrote in its election program: "Our role must not be limited to sending soldiers to hot spots in the world. Poland's participation in restoring stability and peace should not exceed our real capabilities; instead, it should be linked to our political and economic interests" (PO, 2007). These issues were also addressed by Prime Minister Donald Tusk, representing PO, in his exposé:

> The stabilization missions in which we participate are an instrument of Polish foreign policy. (...) Poland's participation in the mission determines our credibility in the Alliance. We must and want to take our allies seriously, we must and want to show them solidarity. After all, we would expect the same from them if Poland found itself in need. Participation in the stabilization mission in Afghanistan also involves creating a barrier

> against the spread and entry of drugs into Europe, including Polish cities and schools. This is also in our direct Polish interest. (Tusk, 2007)

In the next campaign – in 2015, after the annexation of Crimea – the demand to shift the military's role from expeditionary to territorial defense was expressed by all parties. However, it all began with Iraq.

On March 18, 2003, President Aleksander Kwaśniewski and Prime Minister Leszek Miller addressed a joint conference at the Presidential Palace. The President announced that he had received a request for an *Order for the use of the Polish Military Contingent as part of the international coalition forces to contribute to enforcing the Republic of Iraq's implementation of UN Security Council Resolution 1441 of November 8, 2002, and related earlier UN Security Council resolutions* (Kwaśniewski, 2003). He explained his decision in a statement of more than 2,500 words. The key word was "no" – to an expression of disagreement with Iraq's actions, the passivity of other countries, and the threatened level of international security.

The president said:

> We are participants in the great anti-terrorist coalition that was established after the events of September 11, 2001. We remember our pain, including that associated with the death of our compatriots in the rubble of the WTC in New York. We remember the deep conviction we all had at the time that the war on terrorism must be carried out to the very end. Today I want to reiterate these words: the war on terrorism must be carried out as consistently as possible and to the very end, effectively and well for all of us. (Kwaśniewski, 2003)

The president justified his decision a day before the planned start of hostilities, when Polish soldiers, including special unit GROM, were already on Kuwaiti territory, waiting for a signal that operations could begin. The situation was unique because it was the first time Poland was to take part in hostilities, rather than a stabilization operation, and it was taking place without a mandate from an international organization. Hence, the strong words about the extraordinary threat posed by Iraq and the need for such extraordinary action to defeat terrorism and prevent the proliferation of weapons of mass destruction:

> Poland has never favoured war solutions and never will. We are a nation with peace deeply at heart because, in our history, we have too often been the victims of war, partition, misfortune, rape, and terrorism. (...) Today, wiser from that tragic experience, we appeal to the international

> community: peace yes, but not at the price of accepting crime, violence, and terrorism. (...) We agree with the words of Tony Blair, who stated that there are times in history when, to keep the peace, one must take up arms. I am convinced that this decision is the right one – right because of the dangers we must overcome, because of ensuring world peace, and also due to our allied commitments. (...) I trust that our compatriots will understand this difficult historical moment when we should stand together, in solidarity for security and peace in Poland, Europe, and the world. (Kwaśniewski, 2003)

The issue of sending the Polish Armed Forces to Iraq was initially securitized. Saddam Hussein, along with the weapons of mass destruction he was said to possess and his support for terrorist groups, was considered an existential threat to Poland, prompting the decision to take such a step alongside the Americans and the British, contrary to the opinion of most European countries. Appeals for public acceptance were not superfluous; in 2003, 75 per cent of the public opposed the participation of Polish soldiers in operations abroad. There was also no unified support for these operations in parliament. Therefore, when asked by a journalist, Aleksander Kwaśniewski replied:

> If you were to ask me whether I support war or peace, I would answer peace. If you were to ask me whether I think it is necessary to risk the lives of Polish soldiers, even in a limited capacity, I would also have significant doubts. However, if you were to ask me and my compatriots whether we want to ensure Poland's security and freedom from the threat of terrorism, I am convinced that the vast majority would affirm that they desire such security and do not want to live in fear of terrorism. (Kwaśniewski, 2003)

Figure 11 illustrates the securitization process in the case of the Polish Armed Forces' involvement in the Iraq conflict.

A week later, Prime Minister Leszek Miller faced questions and allegations from parliamentarians. Officially, the point of the Sejm session was titled: *Information of the Government on the participation of the Polish military contingent in the international coalition forces to enforce Iraq's compliance with the UN Security Council resolutions*. In fact, it was a discussion of the principles regarding the use of the Polish Armed Forces, the limits of allied loyalty, and Poland's position in NATO. If it had been held under different circumstances (on March 26, 2003, the day of the debate, GROM had already completed several successful operations in the Persian Gulf and at the port of Umm-Kasr),

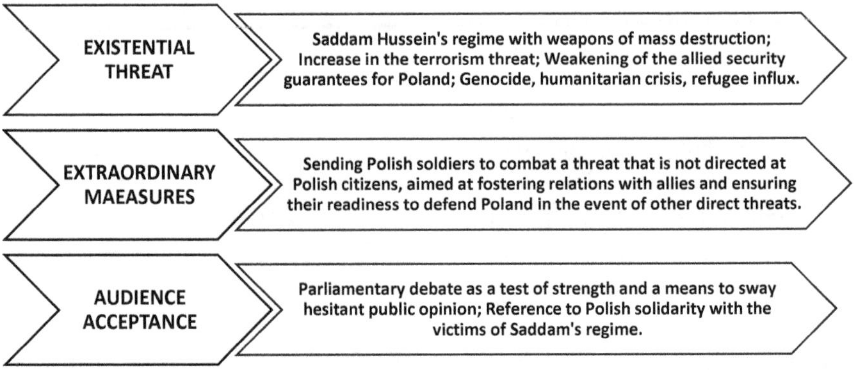

FIGURE 11 Securitization of the participation of Polish soldiers in the operation in Iraq in the narrative of Aleksander Kwaśniewski, Leszek Miller and their supporters
SOURCE: OWN STUDY

it might have led to a cross-party strategy in this key area. However, these circumstances meant that it primarily served as an opportunity to present party positions on an issue that was already a foregone conclusion. The government maintained the position articulated by the president on March 18, and the absence of personal losses appeared to confirm the legitimacy of the measures taken. Nevertheless, in 3138 words, Prime Minister Miller justified the presence of Polish soldiers without a UN mandate in the Middle East.

> How the Iraqi conflict is resolved will determine how the world faces the most serious threats to global security in the 21st century. (…) The Polish government, along with the President of the Republic, made a difficult but justified decision to participate in the operation to disarm Iraq. This decision serves our national interests. We oppose the use of force. (…) However, there are countries ruled by dictators to whom arguments do not resonate. We are a coalition member with the firm belief that the actions taken serve the Polish raison d'être well. (Miller, 2003)

The Prime Minister followed the President's argumentation. He referred to 9/11, the weapons of mass destruction in the hands of dictators, and the possible consequences of further Hussein's activities for Poland:

> Poland must consider terrorism and the threat of weapons of mass destruction as potential threats to our country as well. They can affect us

> directly, and regardless of the form and scale of our support for the allied action in Iraq. We belong to a world that has become a set of interconnected vessels, where remote, quiet and safe places no longer exist.

Leszek Miller pointed out the benefits of Iraq's involvement alongside the US and allied action: "Poland will only be safe in a world where allied commitments are respected." Referring to the responsibility of those present in the Sejm, he appealed for support for the actions that have been started, or at least their acceptance. "We are faced with the necessity of taking one of the most essential decisions fundamental to our security's future. I am deeply convinced that we have made the right decision" (Miller, 2003).

The debate that began divided the room into supporters and opponents of involvement in Iraq. Selected statements are presented in Table 8.

TABLE 8 Selected narratives participants of the parliamentary debate on the participation of the Polish Armed Forces in the operation in Iraq

Arguments from the government side – strong supporters of involvement in Iraq
The SLD is, like society as a whole, peace-oriented. But the SLD also knows that disregarding terrorism is an issue where it is impossible to compromise. Therefore, we believe that since the methods of dialogue failed, it was necessary to reach for other means. We are convinced that the authorities of the Republic acted in the name of the implementation of international law, the implementation of UN resolutions, acted within the framework of a coalition that has only one goal: to disarm Iraq, to bring about the destruction of weapons of mass destruction, to eliminate chemical and biological weapons, to eliminate the breeding ground for future conflicts, future acts of terror. (J. Jaskiernia, SLD)
PSL (coalition partner) doubts
The PSL Parliamentary Club has serious doubts about whether the decision of the President and the Prime Minister was made in accordance with the Constitution. Although it is difficult to find a clear definition of war, after all, the active participation of Polish soldiers in the Iraqi war is not only a violation, it is forcing Iraq to implement Security Council resolutions. (J. Dobrosz, PSL)

TABLE 8 Selected narratives participants of the parliamentary debate (*cont.*)

Opposition arguments for involvement in Iraq
Today, we are faced with a choice between the U.S. policy of forcibly disarming Iraq and dismantling Saddam Hussein's dictatorship, and a policy of continued helplessness and indifference in the face of blatant violations of UN Security Council disarmament resolutions. Poland made the right choice. We sided with democracy and against dictatorship. Now that the struggle against Hussein's regime is underway, it is necessary to maintain a minimum of consensus on Polish foreign and security policy issues. (…) The war against Saddam Hussein provides sufficient justification, justification for our presence there. (B. Komorowski, PO) We have particular reason to believe that the U.S. government made this decision in a situation where international pressure and political leverage had been fully exhausted. One must ask what was the alternative. Was waiting for Saddam Hussein to trigger a much larger international crisis? Was waiting for another aggression, the same as the state led by him undertook against Kuwait, against Iran, against Israel, against the Kurdish community? Or was the alternative a carefree hope that chemical armaments and biological armaments would not become an arsenal preparation for terrorist activities or directly for terrorists? What actions would then be needed? (M. Jurek, PiS)
Opposition arguments against involvement in Iraq
On the night of March 17, 2003, the president, in an authoritarian manner, without agreement and consultation with the main political forces, without the consent of the Sejm of the Republic of Poland, against the will of the vast majority of the nation, made the shameful decision to send Polish troops to the Persian Gulf region to take part in aggressive military action against the Republic of Iraq and the Iraqi people. (…) This war is not our war. (A. Lepper, Samoobrona) That the United States has interests in Iraq. This is their business. We Poles should be neutral. This armed conflict can spread to the entire Middle East. Poland cannot be a party in this war. We have not yet experienced terrorist attacks in Poland, but the president's irresponsible decision may cause the likelihood of their transfer to our territory. (Z. Wrzodak, LPR)

SOURCE: OWN STUDY

The Iraqi issue had to come up in the 2005 campaign, and it did. However, it did not generate as much excitement as it did two years earlier, and it was mainly publicized by opponents of the Polish troops' presence in the Middle East. It was significant that, this time, the issue was not about combat troops as part of Operation Iraqi Freedom, as in March 2003, but rather the Polish Military Contingent, which consisted of 1400 to 1700 soldiers carrying out training and stabilization activities. Most supporters of acting alongside the Americans in the Middle East on this issue remained silent, correctly identifying that the matter evokes ambiguous reactions within the electorate. Only PiS officially declared its support for involvement in Iraq as long as it was deemed necessary. President Lech Kaczyński referred to the presence in Iraq and other volatile regions around the world, albeit briefly, when he assumed command of the military. In 2005, he described the participation of Polish soldiers in foreign operations as a service to humanity, appealing to higher values than merely the security of one's own country:

> For many years, the Polish soldier has been fighting and serving for peace. Today, this effort is seen in Iraq, as well as Afghanistan, Pakistan, the Middle East, and the Balkans; tomorrow, perhaps elsewhere. This service is a contribution to the international community and to humanity. But it is also a service to Poland. (Kaczyński, 2005)

More space was devoted to this issue in the 2007 campaign, when it was already clear that withdrawal from Iraq was a matter of months, not years. There were also mentions of the contingent in Afghanistan for the first time, where the numerical commitment had so far been much smaller (about 100 troops) but was increased. The parties split between those calling for an immediate withdrawal (mostly from both Iraq and Afghanistan) and those making it contingent on allied decisions. The withdrawal was addressed in the exposé by Prime Minister Donald Tusk, making the issue the highlight of his speech: "We have decided that 2008 will be the year in which we begin and complete the withdrawal of Polish troops from Iraq. We have made this decision fully aware of its importance to the Poles, to Polish public opinion, to our army, and to our allies" (Tusk, 2007). On assuming the office of president in 2010, Bronisław Komorowski said this about the role of foreign operations for national security: "We understand that for the sake of our own and collective security, we should engage our forces in allied operations. It is important that the objectives, necessary means, and responsibilities for such undertakings be defined through joint reflection" (Komorowski, 2010).

In 2011, only a group of 15 soldiers responsible for training the Iraqi Security Forces remained in Iraq, while the eighth and ninth shifts of PKW Afghanistan, which carried out stabilization and training tasks, each numbered about 2,600 soldiers. Hence, references in the election programs were primarily to the latter mission. The traditionally "pro-expeditionary" PiS presented the necessity of maintaining the contingent in Afghanistan in this way: "Our country's participation in expeditionary actions served to build allied ties and real solidarity among NATO and EU countries. The experience gained during the mission led to groundbreaking changes in matters of equipment and training of our army" (PiS, 2011). The PO, running for re-election, was less enthusiastic. It was aware of the results of public surveys indicating low support for foreign military interventions, so it tended to focus on success, which was to be the "smooth" completion of the mission. The word: "terminate" was the axis around which this part of the program was constructed: "In 2008 we smoothly ended our participation in the Iraq mission. (…) Now we will prepare a so-called road map to complete the Polish mission in Afghanistan" (PO, 2011).

The opposition was no longer as diplomatic. The SLD, which in 2003 convinced parliamentarians and the public that it was in Poland's best interests to participate in the Iraq operation, was now pointing out that operations in Afghanistan were ineffective and should be withdrawn. "In 2012, the Afghan mission should be completely finished" (SLD 2011). Other opposition groups – Samoobrona and Ruch Palikota – have spoken out directly against involvement outside the country: "Many NATO member countries do not participate in missions, which does not reflect negatively on their international standing" (Ruch Palikota, 2011).

Not surprisingly, the timetable for troop withdrawal has been the subject of parliamentary discussions in the new Sejm. The debate on May 24, 2012, following the government's information on the strategy for withdrawing Polish troops from Afghanistan, symbolically closes a nearly 10-year period of significant discussion about the legitimacy of engaging the Polish Armed Forces in operations abroad, not arouse huge emotions because the assumptions presented by the minister aligned with the expectations of the largest opposition groups (a joint exit in 2014, as agreed at the NATO summit in Lisbon). Defense Minister Tomasz Siemoniak did not use high-flown words or elaborate comparisons. He did not justify the longer-than-expected opposition exit from Afghanistan as a Polish *raison d'etat*.

The opposition, along with the coalition's PSL, which is traditionally skeptical of overseas operations, scrutinised the PO from various perspectives for its inconsistency in the announced plans. Questions arose regarding shifting dates, costs, and profits. A representative of the PiS party, who chose to participate in foreign operations as long as the allies expected argued: "We went in together and we will go out together" (M. A. Kamiński, PiS, 2012). His club colleague pointed out that a premature withdrawal could do more harm than good. "In 2008, by a decision of the PO government, we withdrew from Iraq. It was a bad, populist decision. We withdrew at the worst time, when the situation was already actually beginning to stabilize. We left early and did not get any benefit from it" (Jach, PiS, 2012).

PSL faced a dilemma regarding the split they had entered: formally, the party was in the government that authorized remaining in Afghanistan, while privately, a significant portion of representatives opposed further participation in the operation. Consequently, the strategy was to simultaneously highlight the advantages and disadvantages of the current situation:

> On one hand, the Afghan operation is an invaluable contribution to reducing terrorist attacks in Europe, and from the perspective of systemic changes in the Polish army, it has allowed for the accumulation of experience. On the other hand, however, it was purchased at the cost of Polish soldiers' blood and significant financial expenditures. (Górczyński, PSL, 2012)

The SLD spoke most categorically, "forgetting" who initiated the large-scale involvement of Polish soldiers outside the country's borders. "The war in Afghanistan is unwinnable. Of course, terrorism must be fought, but with other methods, using intelligence, special forces and even drones" (Zaborowski, SLD, 2012).

For obvious reasons (the PKW ISAF mission ended in December 2014), these issues did not arise in the 2015 and 2023 elections. PKW Afghanistan, in the new Operation Resolute Support Mission, consisted of between 120 and 320 soldiers undertaking advisory and advisory-training tasks for the Afghan Security Forces. Poland symbolically returned to UN-led missions. However, it became clear that after two costly and undistinguished operations, Poland, like many other European countries, would not opt to re-engage ground forces for many years, at least in Syria. The time has come for a symbolic engagement, manifesting a presence by performing non-combat tasks, both to minimize costs and to protect its own forces.

One of the most hotly debated issues in the global military sphere – armed interventions in third countries – has not escaped Poland and Polish political discourse. Poland's support for the international coalition intervening in Iraq in 2003, without any universally recognized mandate, called for a securitizing act. The SLD, which at the time had both a government majority and a president, faced the daunting task of convincing public opinion in Poland that intervention in a distant country was closely linked to the safety of citizens on the streets of Warsaw. The argument unfolded in multiple directions: it was pointed out that weapons of mass destruction in the hands of an unpredictable dictator could threaten the entire world and cause far greater damage than the 9/11 attacks. Arguments were made about a global world changing under the influence of terrorism, in which it is crucial to combat threats at their inception, away from national borders. While this reasoning was not without merit, at first glance it seemed like a significant stretch, leading to the second, more nuanced line of argument: by intervening in Iraq alongside the US and UK, Poland is establishing itself as a loyal and faithful ally that stronger partners will not abandon in times of need. Thus, while it has not entirely succeeded in depicting Iraq under Hussein as a substantial existential threat to Poland, it has been argued convincingly enough that non-intervention in Iraq undermines our potential protection in situations where Polish territory may be directly threatened. We are a young, new ally who, as it were, must "pay the buy-in," prove our credibility, and "earn" equal treatment.

The existential threat necessitated urgent and extraordinary measures. Consequently, the decision to engage in the operation within days, without consulting parliament, was based solely on the prime minister's and president's decision. At the same time, however, given the uniqueness of the situation, the uncertainty of public reaction, and finally, the costs associated with the intervention, SLD first appeals for understanding and support from other parliamentary groups, and then seeks to evoke Polish society's sense of justice and solidarity with the Iraqi people oppressed by Saddam. This approach should be regarded as successful: after all, SLD lost power in 2005 not because it decided to send troops to Iraq, but due to the scandals within its own structures.

Military intervention in Afghanistan was no longer securitized. There was no reference to an existential threat, no argument about the need for extraordinary measures. It was a continuation of the previous "normal" policy, already desecuritized.

3 Conclusions

Polish political discourse after 1989 addressed classic security issues, such as defense capabilities, particularly the condition of the army. However, these issues were not securitized due to a consensus on the necessity of appropriate spending on armaments, given Poland's geopolitical location. Defense matters were understood uniformly by the largest political parties and accepted by society. An exception to this was the participation of Polish soldiers in the Iraq War, which was viewed differently due to its unique nature and a general reluctance to deploy troops outside the country for combat missions. In this instance, securitization was necessary: referencing an extraordinary threat and exceptional circumstances, including allied commitments, was required to foster fragile support for the expedition. This was not an issue that could be resolved through "normal politics." However, an analysis from a historical perspective indicates that this was unjust securitization: the threat was not existential, neither for Poland nor globally, and not solely due to the lack of evidence that Saddam Hussein's regime possessed weapons of mass destruction.

Military security discourse in the case of Poland indicates that there is no need for mandatory securitization regarding security issues, as long as they relate to accepted and non-controversial aspects. Traditional security sectors, once questioned and discussed, are now mostly a matter of "normal politics."

CHAPTER 4

Political Security Discourse

Aspects of political security have been among the key topics addressed in the political programs and speeches of prime ministers and presidents since 1989. Among the most relevant themes to analyze in this area are:

1. allied security;
2. special relations with other countries, particularly EU countries, the US, and Russia;
3. terrorism as a global and local threat;
4. combating crime and corruption.

As in the case of military security, not all issues were securitized during the 30 years of the Third Republic; some were merely politicized, without necessitating special action or being presented as existential threats. An issue that received considerable attention and was viewed as fundamental to Poland's security – that is, Poland's entry into NATO – did not exhibit the hallmarks of securitization because, as with the development of the Polish Armed Forces, there was complete consensus on it not only among the political elite but also – society. Thus, there was no need for extraordinary measures. On the other hand, regarding integration with the European Union, one can point to attempts at securitization, carried out, for example, through messages sent to opponents of integration, which often referred to the statement that by being outside the Union, Poland risks: isolation, economic exclusion, being cut off from Europe, etc. In the last analyzed campaign from 2023, however, themes about the Union as a threat to Poland's security, especially its sovereignty, emerged. There was a definite securitization of relations with Russia. Finally, there is also securitization concerning terrorism, which was effectively presented as an existential threat to Poland and its citizens. Periodically, certain groups securitize issues of fighting crime, especially organized crime and corruption.

1 Discourse on Allied Security (NATO and EU)

Kornel Morawiecki was the first Polish politician to speak openly about aspirations for NATO membership in 1991, when the country was still part of the Warsaw Pact and "hosted" Russian troops on its territory. He could afford to do so because he did not represent the ruling party. Politicians who had engaged

© PAULINA POLKO, 2026 | DOI:10.1163/9789004745414_006

in talks with Western partners understood that such appeals unnecessarily irritated both sides: Russia, which was reluctant to part with its former empire, and the West, which was in no hurry to accept underinvested, backward armies into its ranks. Even politicians who later took much firmer stances, in the late 1980s and early 1990s, spoke, for example, of Porozumienie Centrum (PC) (1990), "taking the current political situation into account." The first step was to get out of the old alliances and for the Soviet troops to leave Poland. The process of transformation on the European continent should move toward the emergence of a pan-European collective security system. It should proceed with security guarantees for all its participants. Even before the circumstances allowing its dissolution arise, the Warsaw Pact should be adapted to the new political situation (PC, 1990).

When the RWPG and the Warsaw Pact were dissolved, and the agenda for the withdrawal of Russian troops from Poland began to take shape, demands for establishing a new order began to emerge. These ranged from general proposals to those calling for outright cooperation with NATO (and the EU) for future integration. "Poland should base its security on developing European structures for cooperation and integration" (SdRP, 1991). "We advocate for overcoming the continent's division and for Poland's active participation in shaping the European security system. Poland should withdraw from the remnants of the Eastern Bloc's political, economic, and military structures" (UD, 1991).

The most broadly cautious approach to integration, when it was realistically not possible in the short term, was presented by the UD program. Its leaders acknowledged the limitations mentioned earlier. Therefore, the proposal was to first apply for accession to the CSCE rather than NATO and the EU.

> In considering Poland's security, we should go beyond traditional factors involving relations with two powerful neighbors. A clear answer to the question of the future system of international security, capable of permanently guaranteeing Poland's independence, remains difficult to articulate today. The response to current threats, challenges, and opportunities must involve a dynamic national development policy and the active construction of a complex, multilateral system of international ties. This new system of linkages should enable us to achieve, in the future, a level of security comparable to that of the democratic states of Western Europe. (UD, 1991)

Some parties called for integration into NATO. However, these parties did not significantly influence the government's structure or foreign policy, as previously mentioned. "The only real force standing guard over Europe is the North

Atlantic Pact. In this situation, we must establish closer cooperation with NATO" (PSL Solidarność, 1991). "In the interim period, Poland's security should be based on bilateral agreements with both NATO and the USSR, as well as multilateral agreements with the countries of Central and Eastern Europe" (Solidarność Pracy, 1991). "For Poland's security, we must seek political ties and cooperation with Western countries more explicitly than before" (UD, 1991).

The positions of prime ministers and presidents at that time were similarly conservative. Tadeusz Mazowiecki (1989) noted clearly:

> Poland wants to co-create a new Europe and overcome the post-war divisions of the world. (...) Our opening to Europe, however, does not mean rejecting our previous ties and commitments. If we reiterate today that the new government will respect Poland's allied commitments, this is not a tactical appeasement ploy; it flows from our understanding of the Polish raison d'état and analysis of the international situation. If the day comes when European security does not require military blocs, we will bid them farewell without regret. (Mazowiecki, 1989)

Two years later, Jan Krzysztof Bielecki (1991) could already afford to begin his exposé by expressing joy at the collapse of the RWPG and the Warsaw Pact. In 1992, Jan Olszewski went a step further in expressing Poland's integration aspirations: "We aspire to a Poland that will be a full member state of the European Communities. This connection will bind our democracy to others and ensure the country's security" (Olszewski, 1992). "The most crucial matter today is Poland's integration into the EEC," said Waldemar Pawlak (1992). "My government will strive to make Poland's security equal to that of other European countries, linking this to bringing the prospect of Poland's membership in the North Atlantic Alliance closer," declared Hanna Suchocka (1992).

In the 1993 elections, as the last train cars carrying Russian soldiers departed the country, declarations about new alliances became more resolute. Right-wing, post-Solidarity parties spoke somewhat openly about NATO and the Union, while left-wing parties discussed a new order that combined the needs and approaches of East and West. "We advocate an immediate and vigorous effort to admit Poland into the Atlantic Alliance. We need to feel secure and break the dependencies that fetter us" (PC, 1993). "Poland's defense doctrine should be based on close cooperation leading to a bonded alliance with NATO member states constituting a guarantee of state security" (Partia Chrześcijańskich Demokratów, 1993). "The main direction of our search for development and security opportunities connects us to the European Community and NATO"

(UD, 1993). "Poland's security should be sought in the future collective security system in Europe" (SLD, 1993).

Standing at the head of the government again, W. Pawlak (1993) declared openly: "integration with the European Union, closer cooperation, and future membership in NATO and Western Europe Union." Such declarations will be a regular feature of speeches from now on. "I understand Poland's membership in the European Union, the Western European Union and NATO as part of expanding the zone of security and stability," Józef Oleksy said in 1995. "Poland's interests lie primarily in becoming part of the European Union and the North Atlantic Alliance as soon as possible (...) Our membership in NATO is not directed against any country or group of countries" (Cimoszewicz, 1996). Aleksander Kwaśniewski (1995) said in his speech to the National Assembly simply: "external security is first and foremost Polish participation, Polish presence, Polish membership in the North Atlantic Alliance."

The excerpts from Prime Ministers' speeches quoted above effectively illustrate the style of the debate on NATO membership in the 1990s. It fell under the security category, but the narrative lacked references to public acceptance or the need for extraordinary measures. This was due to the general consensus among the main political factions on the direction of integration. There was no necessity to seek public support or propose non-standard measures. Thus, the focus was on the technical aspects: the roadmap for integration, cooperation under currently feasible terms, or the modernization of the army. The issue of membership in Western organizations was presented as a significant, key, strategic element of Polish state policy, the best way to ensure regional and global security, but was placed in the realm of normal politics. Securitization occurred in the 21st century, first when the requirements for bringing Polish Armed Forces to NATO standards were analyzed, second during involvement in NATO operations abroad, and third when Poland sought the presence of Alliance troops on its territory.

The 1997 parliamentary campaign proceeded with a similar calm conviction about the inevitability and appropriateness of a NATO integration direction. "Poland's external security requires first and foremost that Poland join NATO" (UP, 1997). "Poland's full **integration** into the security structures of democratic countries" (UW 1997). "Ensuring the country's security – initiating the process of integration with European and NATO structures" (SLD, 1997). Crucial for "ensuring Poland's security is its integration into the collective security system in Europe. The PSL supports efforts to bring Poland into NATO" (PSL, 1997).

Standing at the head of the new AWS and UW government, Jerzy Buzek (1997) spoke in the same vein:

> We want to become a full member of the Euro-Atlantic community as soon as possible, which in today's world is the most important guarantee of civilizational development based on the free market and democracy, on the freedom of individuals and nations. Poland in the North Atlantic Treaty Organization is a permanently secure, permanently independent Poland. (Buzek, 1997)

"We joined NATO," Aleksander Kwaśniewski said simply in 2000, taking office as president of Poland for the second time. Becoming a member of the North Atlantic Treaty Organization began a new period of thinking and talking about the Alliance. Since then, the debate has been dominated by themes of (1) Poland's role and fulfillment of obligations, including modernization of the army and the organization of military command, (2) NATO's strategy and its involvement in conflicts outside the area of the Alliance countries, and (3) calls for NATO reform, especially after 2014 and 2022. Hence, the programs of the SLD and PiS in the 2001 elections included extensive passages relating to the first of the themes mentioned above:

> Fulfilling the external and internal commitments of the Polish Armed Forces will require significant expenditures, a major restructuring effort, and mitigation of the social consequences of the reduction in the size of the military and the military infrastructure. (SLD, 2001)

> We have been admitted to the Alliance on credit, so to speak, and without a change in this state of affairs we will not be able to take advantage of the opportunities offered by our geographic location and population potential. (PiS, 2001)

The requirement to "repay the loan" didn't create securitization circumstances – although it was necessary to meet the high standards and level of training, investing considerable money, and changing the mentality, habits, and mindset of the command staff in particular. These measures were accepted, as it were, by acclamation, with no significant dissenting voices. No "extraordinarity" was needed. When taking over the prime minister's chair, Leszek Miller (2001) spoke in general terms: "security is not given once and for all. Of fundamental importance in this regard is our membership in the Atlantic Alliance." His successor, Marek Belka (2004), standing before the Sejm at the time of the presentation of his government, said this about interaction with NATO: "We are fulfilling our commitments in Iraq, supporting efforts to engage NATO and for a speedy transfer of power to the Iraqi people" (Belka, 2004). Iraq was a significant test for the Alliance countries, whose ultimate involvement in the region was far less than that of the US and its coalition partners. It also showed that the Allies were not unanimous on many issues, and Poland had to find a place for itself in this network of relationships. Hence the proposals in the following years for closer relations either with the US or with European countries, or possibly calls to keep an "equal distance." These topics, as well as the issue of modernizing armed forces, were raised extensively in the 2005 PiS election program. "We will consolidate our place in the North Atlantic Alliance as a contributor state to the global security architecture. Our goal is to ensure Poland's proper role in NATO" (PiS, 2005). Not all parties shared this view. SLD staked its position in this regard on relations with the European part of NATO, through the EU: "We are for strengthening the European Union. In favor of a common foreign and security policy" (SLD, 2005).

In his exposé, Prime Minister Kazimierz Marcinkiewicz heralded symmetry in relations, with Poland in the role of peacemaker-connector of "both sides of the Atlantic":

> Integration within the structures of the European Union is a guarantor of development and prosperity, and the alliance with the United States within NATO is a guarantor of security. The two options, European and Atlantic, should not compete with each other. (Marcinkiewicz, 2005)

This symmetry was no longer evident in his successor. Jarosław Kaczyński (2006) put the alliance with the US on a par with NATO membership (J. Kaczyński, 2006).

The campaign prior to the 2007 elections emphasized the need to strengthen defense at the Union level, address changes in NATO, and refine the approach to foreign operations. Even PiS acknowledged the necessity of enhancing cooperation on the continent. "We aim to strengthen NATO while developing a unified European defense policy" (PiS, 2007). PO, while not belittling the role of NATO (it called the Alliance "the main guarantor of Poland's security"), advocated strengthening EU security policy: "It is the EU and the North Atlantic Alliance that bear responsibility for global peace, security and stability of the international order" (PO, 2007). The PO's symmetrism will be the party's enduring approach to global and regional security policy. The PSL had a similar approach: "We will ensure Poland's international security through participation in NATO, the UN, EU, OSCE structures" (PLS, 2007). The Left, on the other hand, saw European integration as the first pillar of the country's security:

> Increasing Poland's security, strengthening its sovereignty and authority in Europe and the world through activity and initiative in the process of deepening European integration, strengthening alliances, and building good-neighborly relations and regional cooperation. (LiD, 2007)

Prime Minister Donald Tusk left no doubt about the North Atlantic Alliance's leadership in providing regional security, but he also acknowledged the EU's role. For the first time at such a high level, a candid statement was made that military involvement outside the country's borders is not due to a threat posed by the intervening country or humanitarian motives, but rather to Allied commitments, which also reflects Poland's future potential needs regarding the defense of its borders. "We view our contribution to NATO's expeditionary missions as an investment that the Alliance will reciprocate with a greater commitment to security at NATO's external borders" (Tusk, 2007). The already-quoted excerpt from Bronisław Komorowski's 2010 speech carried the same content that the prime minister had previously conveyed. "As a member of NATO, Poland is aware of its allied obligations. We understand that for the sake of our own and collective security we should engage our forces in allied activities" (Komorowski, 2010).

The 2011 campaign was full of references to allied relations and was dominated by themes of (1) institutional reform, (2) strengthening Poland's position, and (3) criticism and disillusionment with allied organizations. PiS dedicated the most attention to NATO in its program, beginning with the understanding that allied relations alone do not guarantee complete security. This new narrative, stemming from Russia's attack on Georgia, elevated Poland-NATO relations to a new stage, which culminated, by the way, in the realization of the postulate. The party extensively addressed the future of the organizations in question and their potential to ensure security. For the first time since the Alliance's involvement in foreign operations, the disparity between the interests of the countries for which they are a priority and those of Poland, which has legitimate security concerns about its territory, was clearly highlighted. There was a theme of "reciprocity." "Poland should make its support for efforts to make the Alliance a vehicle for expeditionary operations contingent on recognition of our demands to maintain effective collective defense mechanisms" (PiS 2011).

The PO similarly supported the view that membership in security organizations does not solve all problems: "Being a full member of the strongest military alliance in history – NATO – Poland must take into account various threats in its defense policy" (PO, 2011). The Left put security provided in the European system first. In the case of NATO, SLD was betting on: tightening, deepening cooperation, including participation in more potential missions abroad. PSL called for a combination of installing NATO infrastructure and deploying troops on Polish territory, while maintaining good relations with Russia. The program offered no practical guidance on how to achieve this. Two other parties running in these elections – Ruch Palikota i Samoobrona – showed less support for NATO and its policies than the mainstream parties. "Samoobrona is not opposed to Poland's presence in NATO. However, the alliance should serve mainly to develop the Polish army and ensure the defense of the Republic, not to participate in further senseless wars" (Samoobrona, 2011).

Donald Tusk's second exposé was shorter than the first and mostly devoted to economic issues. There was not a word about NATO in it. Ewa Kopacz, who replaced him in 2014, referred to the Alliance in one sentence: "The Ukrainian crisis shows how important it is that we do not have to face this challenge alone, because the European Union and NATO are behind us." But back in the 2015 parliamentary campaign, the PO spoke the language of the PiS:

> We will increase the military presence of NATO and the United States on Polish territory. (...) Participation in the North Atlantic Alliance will remain a key element of Poland's military security. Our goal is to convince the allies to establish permanent military bases in Poland, confirming Poland's special position in the Alliance. (PO, 2015)

This special place of Poland in NATO, achieved through previous loyalty to key Alliance countries, has appeared multiple times in both the PO and PiS narratives. Gained through involvement in Iraq and Afghanistan, as well as the purchases of the F-16/F-35 or the Patriot system, it became a crucial measure to eliminate the threat emanating from Russia. After 2014 and 2022, the issue of protecting Polish territory from the threat from the East began to be securitized. To defeat this threat, it was postulated: increase defense spending, change the approach to the use of the Armed Forces, establish the Territorial Defense Forces, "pull" NATO to the territory of Poland.

Once again, the most comprehensive program on collective defense was presented by PiS. Party began with a diagnosis that included criticism of NATO's insufficient guarantees of assistance in the event of an emergency. "We do not expect privileges. We want to equalize our security status with that of Western Europe. (...) Poland cannot be a second-class member in NATO" (PiS, 2015). "Second-class" membership, demanding equal treatment with "old" NATO countries, is a new element in the discourse on the role of the North Atlantic Treaty Organization. After 2014, as the conviction began to grow in some political circles in Poland that a conventional attack on the country's territory by

Russia was not only possible but highly probable, demands that NATO and the US pay their debts for Afghanistan and Iraq emerged not only from Pis. In addition to investing in its own army and limiting its use abroad, the demand for NATO military and infrastructure presence in the former Warsaw Pact countries served as a complementary element to extraordinary measures aimed at countering the Russian threat. All these elements were incorporated into the PiS program.

> We naturally view NATO's role through the prism of our national security interests. (…) Poland needs real – i.e., material – security guarantees and the strengthening of NATO's presence not only in our country, but also on the territory of other members of the Alliance from Central and Eastern Europe. (PiS, 2015)

"The fundamental goal for the coming years remains unchanged – the North Atlantic Alliance must continue to be a politically and militarily credible guarantor of the security of NATO countries," Beata Szydło said when taking over the Prime Minister's portfolio (2015). The tone of Andrzej Duda's statement was similar (2015, a month later): "The task for the near future is to strengthen allied guarantees in the North Atlantic Alliance. And here the matter must be put unequivocally. We need greater guarantees from NATO" (Duda 2015). Two years later, Prime Minister Mateusz Morawiecki (2017) argued that the ambitious plans of 2015 are already largely in place, and his main concern now is harmonizing EU and NATO cooperation, where Poland has a significant role to play. "NATO is and will remain the foundation of our security, while the United States is our main ally" (Morawiecki, 2017).

The 2019 parliamentary campaign maintained the trends outlined in 2015. PO, running as part of the KO, reiterated the need for balance in Europe's involvement in the security of Poland and the region, while maintaining NATO's leading role. "We will support any form of cooperation within the EU that develops the defense capabilities of member states in coherence with NATO" (KO, 2019). Koalicja Polska (KP), composed of PSL and Kukiz15, spoke in a similar tone. The direction of basing security on the EU and NATO was obvious, although the order of the mentioned institutions – significant. "The pillar of our security is firmly embedded in the structures of the European Union and NATO. Our foreign policy cannot be limited to courting one country" (KP, 2019).

PiS (in the Zjednoczona Prawica, ZP coalition) has upheld the line of basing Poland's security on NATO/US, with the possibility of including the European Union.

> Poland and the United States have become important partners in strengthening security, stability and development in Europe and the world. Poland has become the security keystone of our region and the entire continent. (…) We will strengthen NATO's eastern flank by expanding the Alliance's defense infrastructure of strategic importance on the territory of our country. (ZP, 2019)

Prime Minister-designate Mateusz Morawiecki (2019) reiterated the main tenets of his party's election program: "We will defend the alliance of Europe and the United States. Just as the European Union is the guarantor of the continental order, NATO takes care of the global order" (Morawiecki, 2019). Andrzej Duda's speech (2020) was also not surprising: "We must work to strengthen Euro-Atlantic ties and the North Atlantic Alliance" (Duda, 2020).

Allied security returned with redoubled force in the 2023 election campaign. The consensus of the main parties on the inevitability of relying on collective defense was almost total: "Embedding in NATO (…) Strong Poland in a Strong Europe" (Lewica, 2023); "Strong in Alliances: Poland will become the security foundation of NATO's eastern flank. Thanks to this, Poland's role as a strong partner will also be important for our allies" (PiS, 2023). Only Konfederacja (2023) pointed out: "successive governments have wrongly seen the only guarantee of Poland's security – that is, participation in alliances and international organizations," but NATO membership itself was not questioned. The importance of the Alliances was also discussed by the two prime ministers delivering the 2023 exposé: "We are responsible for the security of a key section of the eastern flank of NATO and the European Union" (Morawiecki, 2023); "Poland is and will be a key, strong, sovereign link of the North Atlantic Alliance (…) I ask everyone to stop pretending that the threat to Poland is our friends and allies of the North Atlantic Treaty Organization and the European Union" (Tusk, 2023).

•••

The North Atlantic Treaty Organization is the primary alliance regarded as the guarantor of Poland's security. This perception exists both during the period when its membership was not yet openly discussed (early 1990s) and at the time of association, culminating in full membership. Since February 2022, following the Russian invasion of Ukraine, it has been under special review. The views regarding the organization, Poland's role within it, and the expectations from the Alliance have been evolving. The prevailing narrative at the beginning was that we joined NATO "on credit," needed to increase military spending and modernize forces, and had to "earn full membership" through participation in overseas operations. This shifted by the end of the first decade of the new century, influenced by the events in Georgia (2008) and Ukraine (2014 and 2022). In the second decade of the 21st century, Poland, representing other Central and Eastern European countries, began demanding that NATO and its leaders withdraw from the 1997 agreements with Russia and deploy permanent military components and infrastructure on the territory of new alliance members. Finally, a strong response to Russian aggression in 2022 included providing military support to the struggling Ukraine.

NATO membership, Poland's position in the Alliance, and the level of guaranteed support in the event of an external threat are issues that were not themselves subject to securitization but intersected with related areas. Being a member of this organization, based on terms formulated by Poland, is a guarantee of security. Therefore, it was necessary to first modernize the army and align it with NATO standards, then to engage in foreign operations whose connection to a direct threat to Poland is difficult to establish, and finally, to secure the permanent presence of allied troops on Polish territory, even at the cost of building infrastructure from our own budget or purchasing equipment from allies.

2 Discourse on Special Relations with Other Countries and Regions (Europe, USA, Russia)

2.1 *Europe*

In the initial years following the first partially free elections of June 1989, enthusiasm for integration into the Western world was moderate. Several factors influenced this perspective. Solidarity's electoral victory marked a significant change in Poland, but not in its surrounding environment. The People's Republic of Poland (a name that remained until December 1989) was still part of the Warsaw Pact, and the constitution, amended in 1976, contained a provision for friendship with the USSR. Soviet troops were stationed on Polish territory, and, apart from radicals, few advocated for accelerating relations with other countries. A key aspect of relations with nations west of Poland was the maintenance of the Oder-Neisse border following German reunification. Consequently, most discussions at that time centered on peace and finding a place within the European family, emphasizing an atmosphere of friendship and cooperation while respecting autonomy. Between 1989 and 1991, specific references to deeper integration were noticeably absent. Instead, the focus was on dissolving old divisions rather than establishing new blocs.

The first prime ministers spoke in a similar vein. Tadeusz Mazowiecki (1989) declared his desire to "co-create a new Europe and overcome the post-war divisions of the world, which are now completely anachronistic," but in the next paragraph he made it clear that our "opening to Europe does not mean rejecting existing ties and commitments." He was looking forward to "a time when European security would not require military blocs," but it did not appear from the statement that this would happen soon. He talked a lot and in a decidedly non-confrontational way about the USSR, and devoted more space to it than to the West or Europe. Of the European countries, he mentioned Germany (West Germany and East Germany), emphasizing the need to develop good relations.

His successor, Jan Krzysztof Bielecki (1991), expected from the West above all *a radical reduction of our debt.* He spoke cautiously, though less conservatively than his predecessor, about international relations: "We will base our security on good relations with neighboring and other countries in the region, on the emerging pan-European security system" (Bielecki, 1991). Jan Olszewski (1991), who delivered his exposé in the same calendar year, but less than 12 months apart, was already more radical in his statements.

> "We are today between two worlds: the Western world, which is peacefully forming and uniting, and the Eastern world, which is disintegrating. We do not yet belong to the former, already – to the latter. We have decided to join the structures of the West, but for the time being we are without alliances, anchoring, securing. We must make an effort to change this risky state. We are striving for a Poland that will be a full member state of the European Communities" (Olszewski, 1991). *We are convinced that our country is capable of influencing its international environment in a way that corresponds to our interests and at the same time is conducive to the creation of a* ***better, safe and prosperous Europe*** said Hanna Suchocka (1992).

Caution also characterized the first presidents' speeches. In Wojciech Jaruzelski's speech (1989) there was no reference to other countries. Lech Wałęsa spoke of Poland "wanting to be part of a peaceful order in Europe" and a good neighbor, including to Germany, "in which we want to know a friendly gateway to Europe" (Wałęsa, 1990).

In the 1993 campaign, Europe ceased to be seen as a collection of separate countries; it took on the organizational form of the EEC. This trend will continue until membership in the organization is achieved. Individual countries, apart from the US, including Ukraine and a group of Baltic States (possibly the Weimar Triangle or the Visegrad Group), will not be featured in programs as reference points in the security context. Additionally, prime ministers, who traditionally mention foreign partners in their exposés, will do so infrequently. It can be noted that since Poland was given the green light to declare its official integration intentions and was subsequently admitted to NATO and the EU, consciously or not, there has been a shift away from basing security on relations with specific states or smaller groups. The trend of "thinking organizationally" was already evident in Waldemar Pawlak's second exposé: "integration with the European Union, closer cooperation and active expansion of bilateral relations and regional cooperation." He mentioned Germany, the Visegrad Group and the Baltic States, but in the context of economic relations and "general friendliness." Jozef Oleksy (1995) spoke in a similar vein: "Our intention is to continue the strategic line of foreign policy; the line of European and Atlantic orientation and the pursuit of friendly multilateral and bilateral relations."

Włodzimierz Cimoszewicz said bluntly: "We will attach great importance to bilateral relations with other member states of the North Atlantic Treaty Organization and the European Union" (Cimoszewicz, 1996). A year later, Jerzy Buzek (1997) mentioned NATO and the EU, along with integration with these organizations in the context of security. A little over a month after the 9/11 attacks, Leszek Miller expressed the importance of bilateral relations, though without specifics. Marek Belka, who succeeded Miller, focused on economic issues and unemployment. In terms of relations with other countries, he did not extend beyond the "EU membership scheme." The 2005 campaign, like the previous two, did not address the discussed themes. In Prime Minister Kazimierz Marcinkiewicz's exposé, there were general references to the EU, NATO, and support for Ukraine. Jarosław Kaczyński, taking control a year later, similarly to his predecessors, referred to NATO and the Union, with the latter being expanded to include Ukraine.

A little more than six months before his brother's exposé, during his swearing-in as president, Lech Kaczyński spoke of new energy in foreign policy, albeit based on transatlantic relations and the Union. In the case of the latter, however, he recognized the nuances of bilateral relations: the improvement in relations with Germany and France is a positive development, "which does not mean that there are not still, especially in relations with the former country, Germany, very significant problems." He went on to mention Ukraine and the Baltic States, as well as Belarus. He mentioned the Czech Republic, Slovakia, Hungary and the Scandinavian countries, but without any specific context. The overall message was: relations should be good.

In the 2007 elections, the left returned to the policy of distinguishing between EU states as separate entities and not just partners in the organization:

> Germany in a strong and successfully unifying Europe will not be a threat to Poland. On the other hand, Germany, operating in Europe's increasingly weak and shaky international environment, caused by the erosion of EU and NATO structures and mechanisms, could be a serious challenge for Polish foreign policy. (LiD, 2007)

Other parties did not address this topic. The exposé of new Prime Minister Donald Tusk devoted a great deal of space to the cohesion and strength of the Union. The prime minister did not speak about its members except to refer to the Visegrad Group and the Baltic Sea countries in the economic and political context.

Bronisław Komorowski (2010) laconically referred to other than intra-organizational relations with other European countries. "In order to express Poland's European rootedness, I will make my first foreign visit to Brussels, Paris and Berlin. (…) I will also, of course, work to deepen cooperation with other EU partners" (Komorowski, 2010). In the 2015 campaign, the PO was the only one to mention non-EU relations with European countries.

> We recognize the fundamental importance that the Franco-German tandem has for European integration. We want to continue to implement the cooperation with Germany and France within the Weimar Triangle, which was resumed under the PO government, maintaining an open dialogue based on partnership and trust. (PO, 2015)

President Andrzej Duda (2015) emphasized without going into details that good neighborly relations are important "because they strengthen security" (Duda, 2015). In 2020, he referred to the idea of the Trilateral Initiative, which he promoted.

In 2019, the Koalicja Obywatelska, in a similar tone to the earlier PO, declared stronger cooperation with France and Germany. However, it stressed that it was taking place within the EU.

> We will return to close collaboration between the EU's largest countries, including especially Germany and France, and for the countries of the region we will become a worthy representative and valued ally. In this way, we will be a natural link between the largest and smaller countries, Western and Central Europe, the Baltic and the Danube. (KO, 2019)

Koalicja Polska (PSL and Kukiz15) spoke in a similar vein. Zjednoczona Lewica did not address this topic, nor did the Konfederacja (it emphasizes its anti-Unionism instead).

In describing its foreign policy, Prawo i Sprawiedliwość focused not so much on Poland's relations with other countries as on the country's position. The program was constructed in a trend familiar to voters since 2015: politics of dignity and getting up from the knees, resisting the decisions of both the EU as a whole and its individual states. The example of Germany well demonstrated this: "We want our relations with Germany – a very important neighbor of Poland – to be based on the principle of equal partnership, respect for our interests in the region and on the continent, and consistent dialogue." This narrative was also maintained in 2023, where it declared further "rejection of clientelism – swimming in the mainstream and bidding Polish affairs higher" (PiS, 2023). The topic of relations with other European countries didn't appear in the exposés of Prime Ministers Morawiecki and Tusk, other than declarations about supporting the embattled Ukraine in the context of Poland's military security.

2.2 *United States of America*

Despite Poland being a member of the Alliance since 1999, the special relationship with the US has not only retained its importance but has strengthened. In 2020 and 2023, the issue of a strategic alliance with this superpower was even more prominent than ever before, particularly during the negotiation of terms for installing elements of the so-called missile shield in Poland and the permanent presence of NATO troops on Polish territory. Moreover, over this period, there were times when Polish-American relations were heavily securitized.

The first programs of political parties in 1989 and 1990 did not mention the US. Tadeusz Mazowiecki (1989) spoke only of "catching up with the EEC and the United States," primarily in the economic context. His successor, Jan Krzysztof Bielecki, attached "great importance to the development of relations with the US and Canada," but mentioned them in a string of other groups of countries from all continents, without giving special importance. The United States was not the first point of reference in the early 1990s – relations with its closest neighbors were key.

In the 1991 campaign, the US was mentioned sporadically, in several election programs, always in the same context – the importance of good relations, among other security aspects. The importance of US involvement in Europe was presented in general terms – the security of the entire continent. "For the processes of Europe and the Western world, the active participation of the

United States is still important. The country still seriously determines the security and stability of our continent" (KLD 1991). "The contribution of the U.S. to the maintenance of European equilibrium – among other things, through its role in NATO – is and will continue to be very important" (Ruch Obywatelski – Akcja Demokratyczna, 1991). The new prime minister – Jan Olszewski (1991) – was the first to point out that "the US military presence in Europe is a factor of stability." Like his predecessors, he also saw the United States as an economic partner. W. Pawlak saw the US as a stabilizer of Europe's security: "we see their military presence in Europe and participation in European cooperation as an all-around stabilizing factor" (Pawlak 1993). Hanna Suchocka held the line of her predecessors: "Pan-European stability is also guaranteed by the political and military presence of the United States in Europe" (Suchocka, 1993).

In the 1993 elections, only one party referred directly to the North American superpower in security terms: "A continued U.S. presence in Europe remains important for Poland's security" (Ruch dla Rzeczpospolitej, 1993). Others focused primarily on NATO and the possibility of Poland's association with the North Atlantic Treaty Organization as a whole. New-old Prime Minister W. Pawlak (1993), in his second exposé, upheld integration declarations, not deeming it necessary to mention the US separately in the context of security. His successor, J. Oleksy (1995), neutrally indicated that "he would give special attention to the development of political, economic and cultural relations with the United States and Russia." President Aleksander Kwaśniewski, elected in 1995, did not mention the US even once. But Prime Minister W. Cimoszewicz (1996) returned to involving the US in the security of the continent: "The future of Europe depends to a large extent on the involvement of the United States in the process of development and security of our continent. Our relations with the US have essentially taken on an alliance character" (Cimoszewicz, 1996).

The 1997 parliamentary campaign did not note the role of the US in building security. Only one party picked up this theme: "Recognizing the priority of Poland's security, we will participate in building an effective system of collective security in Europe, respecting the special role of the US in shaping the global security system" (PSL, 1997). The new prime minister, representing AWS Jerzy Buzek, used the words strategic partnership in the US context for the first time. "We believe that the strategic interest of both Poland and the whole of Europe is to maintain the American military and political presence on our continent" (Buzek, 1997). However, this trend – common in the 21st century in the perception of the United States by Polish politicians – although started by leaders of the Left (A. Kwaśniewski and L. Miller) will be more strongly represented by parties and politicians of the right, with each year more and more focused on the US, and moving away from the European Union. The 2005 program of

PiS is a good example: "We will work to strengthen the strategic partnership with the United States, which is our key ally and main guarantor of security. Cooperation with the US strengthens Poland's position in the EU" (PiS, 2005).

As can be seen from the above quote, the relationship with the US was to be "strategic, special, privileged, formal, beneficial and deep," but not to weaken joint defense within NATO. The beginnings of the party's narrative, which highlighted the need for a strong bilateral relationship with the North American superpower amid a weakening NATO and a Union that is insufficiently focused on security and defense issues, appeared here. These special relations (extraordinary measures) were then justified by extraordinary threats (possible energy blackmail from Russia, or even an invasion, similar to those in Georgia and Ukraine, and finally – the 2022 invasion). They justified spending on the presence of American troops in Poland and the purchase of military equipment from the US. However, the basis for this undeniable securitization process should be sought as early as the 2005 announcements. Prime Minister Kazimierz Marcinkiewicz (2005) spoke in a similar vein, noting in particular that the EU and the US are not competitors, as they share common threats: "The current challenges of world politics, such as globalization and terrorism, impose a close alliance between the European Union and the US. It is an essential condition for global stability." A strong alliance with a partner from across the Atlantic was also heralded by Lech Kaczyński (2005): "relations with the United States must, in the best national interest, be strengthened, and in difficult moments be maintained, while firmly placing our demands." Six months later, taking over the Prime Minister's portfolio, Jarosław Kaczyński, in other words, directed a very similar message to the MPs gathered in the Sejm: "Poland is in an alliance with the United States. This alliance is maintained and strengthened by joint ventures. These are controversial and not easy undertakings" (Kaczyński, 2006). In the statements of both leading PiS politicians, one can see the "conviction" of courting special relations, the need to take on challenges that are not always popular (this is, of course, about the presence in Iraq), but they are still subject to certain caveats.

Three parties addressed relations with the US in the 2007 elections campaign. PiS again devoted the most space to the issue, again advocating the strengthening of relations, but not without conditions. "Installing elements of a missile defense system in Poland has the dual significance of enhancing Poland's security and strengthening the strategic partnership with the US" (PiS, 2007).

The US was mentioned laconically in their election manifestos by PO and LiD. According to PO, the relationship with the superpower should be: "Allied and partnership, that is, closely linked to NATO and EU security policy"

(PO, 2007). "In matters of foreign and security policy, relations with the United States have traditionally been important for Poland," Lewica asserted (LiD, 2007). The incoming Prime Minister, Donald Tusk, said in his exposé: "With the United States, as with the European Union, we are linked by a community of values. (…) We will try to convince our American partners that our alliance should find expression in a greater American presence in Poland" (Tusk, 2007).

In this statement one can see outlined only a trend emerging in the later policy of Platforma Obywatelska to strengthen relations with the US, not to gain distance from the EU and to support rather than criticize its security policy. In practice, this was reflected in the involvement in operations abroad under the EU flag and in the push for the withdrawal of Polish troops from Iraq. Bronislaw Komorowski (2010) did not give special importance to relations with the US – he saw them in the context of joint NATO membership.

Protracted negotiations over the deployment of elements of the missile shield in Poland and the withdrawal of troops from Iraq earlier than the ally wanted, weakened ties. Hence, in the 2011 campaign, PiS called for a return to the tracks of strategic cooperation: "It is important to take steps to renew our relations with the United States. This will likely require redefining our country's engagement in various directions in such so it involves greater U.S. involvement in our part of Europe" (PiS, 2011). For its part, the PO, which is seeking re-election, did not plan for change. "In relations with the US, we try to present a reciprocity-based policy. Within its framework, we have negotiated good agreements on deploying elements of the missile defense system and on the status of US troops in Poland." It also pointed to the need to seek other than defense benefits from cooperation (PO, 2011). Samoobrona, on the other hand, criticized the presence of foreign military installations on Polish territory and the purchase of American technology.

The moment that restored the pro-Americanism of Polish foreign and defense policy and lowered criticism of the terms of bilateral cooperation was the seizure of Crimea by the Russian Federation in 2014. The weakness of NATO and the EU in the face of this violation of international law showed that in order to realize the demand for a permanent presence of the Alliance's troops in Poland, it is necessary to cooperate more closely with the Americans in this regard. Hence the exposé by Ewa Kopacz, representing the still skeptical PO recently, included an unequivocal statement: "In view of what is happening in Ukraine, it becomes all the more important to strengthen relations with the United States, so my government will make every effort to ensure that the United States increases its military presence in Poland" (Kopacz, 2014). It was a time when the two main parties of the Polish political scene – PO and PiS – spoke with one voice: "The political, economic and military alliance with the United States is a key element in the security of not only Poland, but all of Europe. We will strive to increase the US presence in our part of Europe, first and foremost in the security dimension" (PiS, 2015). Prime Minister Beata Szydło (2015) spoke in a similar tone: "We attach special importance in security matters to relations with the United States of America, which is today and will remain in the foreseeable future the main guarantor of world peace."

The 2019 parliamentary campaign referred in the American context primarily to the presence of US troops in Poland. However, the thread was taken up by only two main groups. Koalicja Obywatelska declared the strengthening of this presence, but "on a partnership basis." "KO is in favor of strengthening and expanding the presence of US and NATO troops on Polish territory. We will base relations with the US on partnership principles, because America does not expect servility, but professionalism" (KO, 2019). PiS, in turn, reproached PO for loosening relations with the US and naiveté in seeing the European Union (and especially Germany) as a guarantor of Poland's security. "It was in the strategic interest of Poland and the entire region to revitalize Polish-American relations. (…) Poland and the United States have become mutually important partners for strengthening security, stability and development in Europe and the world" (PiS, 2019). President Andrzej Duda (2020) spoke in a similar vein. "We must continue to develop our cooperative capabilities with the allied forces present on our territory, in particular with the US military."

The 2023 campaign had a strong focus on strengthening cooperation with the US in the face of the growing Russian threat and the war in Ukraine. There was virtually no rift here between the country's main political forces, as with the speeches of Prime Ministers M. Morawiecki and D. Tusk.

Special Polish-US relations and their various aspects, including the presence of US troops on Polish territory, were subject to securitization processes at times. The need to strengthen these relations was emphasized, especially after 2014 and 2022, in conjunction with the demand for increased spending on the army. This funding was intended to finance not only the expansion of the Polish military but also the procurement of American equipment and the preparation of infrastructure for the presence of US troops. These high costs (extraordinary measures) had to be accepted because the enemy was at the borders and could threaten Polish territory. NATO and the EU had not proven effective as organizations in providing security. Therefore, it was necessary to rely on bilateral relations with the superpower, which, although costly, were certain – American soldiers came to NATO's eastern flank and are still stationed here, even though they "had the farthest to go." Western European

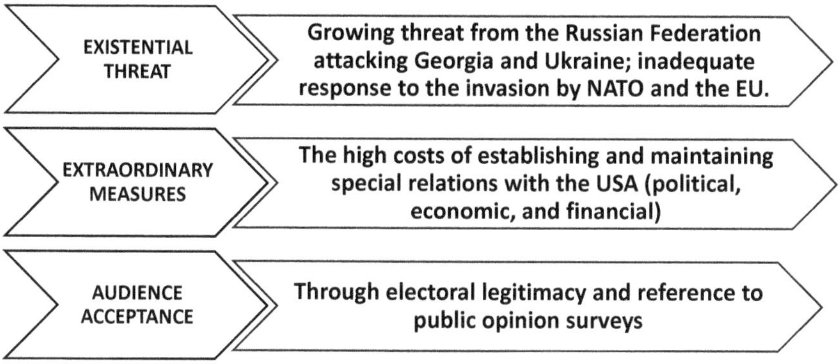

FIGURE 12 Securitization of special relations with the USA
SOURCE: OWN STUDY

countries did not shy away from sending their armed forces near the borders with Belarus and Russia. Many of them, especially Germany and France, had economic ties with Russia, sold it arms, and were not interested in stopping its expansion. We could only count on the US. This narrative, though constructed primarily by PiS, became dominant in the second decade of the 21st century. PiS opponents did not present an equally strong case for strengthening security and defense relations among European countries. Following the 2020 Russian aggression against Ukraine, all major political forces spoke with one voice about the importance of American support. Therefore, the securitized issue of special Polish-American relations was able to undergo a process of desecuritization and return to "normal" politics: the consensus on the inevitability of this path, along with the purchase commitments made and accepted in the US arms industry, meant that there was no longer any reason to call for extraordinary measures. Figure 12 illustrates the process of securitization of Poland-US relations.

2.3 *USSR/Russian Federation*

A neighbour, ally, and partner of Poland, while serving as an overt enemy for at least some of the parties, appeared in political speeches and party programs from the very beginning. The period until 1993 was dominated by themes of loosening relations, leaving the Warsaw Pact, and the withdrawal of Soviet/Russian troops from Polish territory. Thereafter, until the end of the 20th century, the primary issue was NATO expansion and Russia's agreement to this process. In the 21st century, relations were largely defined by energy and, more broadly, economic issues (considered within the security dimension) as well as Vladimir Putin's expansionist policies (Abkhazia, South Ossetia, Donbas, Crimea, Transnistria, all of Ukraine).

In retrospect, there are various assessments of the process of changing alliances after 1989. Many opinions suggest that it could have occurred more swiftly (Strzelczyk, 2002). Years later, however, this contention appears to hold little significance, and the acceleration could be measured in months at most. The Soviet army's combat troops departed Poland in 1992, with the last soldier leaving in 1993. The RP exited the Warsaw Pact with its dissolution in July 1991, while the RWPG did so a month earlier. Considering that until December 1990, the president of Poland was the architect of martial law, and the first fully free elections took place in the fall of 1991, there was significant fragmentation among the post-Solidarity forces, exacerbated by a president who could not unite them, this pace still merits appreciation. It is important to remember that the group (both politicians and their electorate) resisting rapid political change still maintained its strength. Back in November 1989, that is, when Tadeusz Mazowiecki was prime minister, the PZPR proclaimed: "We believe that Poland's alliance with the Soviet Union, based on equal, mutually beneficial state and economic relations, is an important factor in the security and development of our country" (PZPR, 1989). Prime Minister Mazowiecki, in his exposé, asserted that "the government continues to attach great importance to economic relations with the Soviet Union and other RWPG countries." Regarding Poland-Soviet bilateral relations, he spoke diplomatically:

> Mutual relations between a superpower like the Soviet Union and a medium-sized country like Poland are inherently complex. A rational approach to this issue must seek solutions that consider the interests of the great superpower while also respecting our state's sovereignty and its full freedom to shape its internal affairs. (Mazowiecki, 1989)

The caution of the first non-communist prime minister was fully justified. The bipolar world persisted, the PZPR still held power, and a security apparatus remained in place. This was not a time for radical declarations of independence. Fearing a bloody revolution, Mazowiecki advocated for evolutionary changes. Admittedly, Gen. Wojciech Jaruzelski, in his speech following his election as president, did not mention the USSR or bilateral relations, but the newly formed SdRP (1990) made it clear that it would not support a rapid reversal of alliances: "In a divided Europe, it is unacceptable both to extend the NATO zone to the Polish border and to abandon the security guarantees of Warsaw Pact membership" (SdRP, 1990).

There was a realistic attitude among the opposition regarding the possibility of a rapid loosening of the "brotherly alliance," and while the trend toward sovereignty in relations with the USSR and closer ties with Europe was present, it was not radical. In 1990, PC wrote in its program:

> It is in the interests of both Poland and the USSR that the changes taking place respect the right of nations to self-determination, and in a way that could bring the nations of the USSR closer to Europe. Such a transformation would facilitate the creation of a European collective security system to replace the existing blocs in the future. (PC, 1990)

Other parties spoke similarly. Those advocating for a rapid withdrawal from old alliances and a break from the USSR were in the minority. As time passed, with changing opportunities, peaceful democratic reforms in other countries, and the weakening of the Eastern superpower, calls for a change in alliances were proclaimed more forcefully. Taking over the portfolio from T. Mazowiecki in January 1991, Jan Krzysztof Bielecki said in the first sentence of his exposé: "The disintegration of the UW and the RWPG and the withdrawal of Soviet troops in quick succession should complete the work of nations taking their affairs into their own hands." Less than a year later, Jan Olszewski (1991) began his exposé with external threats to Poland still coming from its eastern neighbor. Nevertheless, he declared cooperation with the new order emerging across the eastern border: "Our relations with the Russian Federation, Ukraine and the Republic of Belarus are becoming paramount. Both opportunities and threats are emerging for us in the East" (Olszewski, 1991).

Waldemar Pawlak (1992), in his first exposé, laconically mentioned Russia alongside several other eastern countries with which Poland was to have good neighborly and partnership relations, while noting: "We will see to the implementation of the package of agreements on the withdrawal of Russian troops from Poland." Speaking a few days later, new Prime Minister Hanna Suchocka, like her predecessor, announced: "In relations with Russia, Ukraine, Belarus and Lithuania, we are particularly interested in the following issues: security and balance, economic exchange and the situation of Poles living across our eastern border" (Suchocka, 1993).

In the 1993 campaign, the Russian Federation was mentioned as a neighbor with which Poland maintained good neighborly relations. Russian troops left Polish territory, and the former superpower was temporarily uninterested in expansion. Of course, it still opposed NATO's expansion, but the Alliance itself was not yet prepared for such a step either. After the SLD won the elections, successive prime ministers from the SLD-PSL coalition dedicated little attention to Russia in their speeches and did not present the issue of Polish-Russian relations in a significant way. Pawlak (1993) suggested that relations with all of Poland's eastern neighbors were equally good and unproblematic, while Oleksy (1995) did say: "we are particularly concerned about the development of political, economic and cultural relations with Russia," but mentioned it after the US. W. Cimoszewicz (1996) was the only one of the prime ministers of this term of the Sejm to view Russia in terms of security. When negotiating Polish accession to NATO, he had to face opposition from the Russian Federation.

Nor was Russia a special topic in Alexander Kwaśniewski's speech (1995). He referred to it in common with other Eastern Bloc countries as "neighbors with whom we have shared the good and bad pages of history for centuries." In the subsequent 1997 election campaign, Russia did not appear as a security topic. Politicians focused on NATO integration and no one intended to "tease the bear." Taking over the Prime Minister's portfolio, Jerzy Buzek (1997) declared attaching special importance to relations with Russia. "Unencumbered by the baggage of historical pasts burdening the previous coalition, we will make efforts to make Polish-Russian contacts a partnership and as good as the mutual knowledge of Poles and Russians is." Pointing out areas where this cooperation has a chance to develop, he referred not to security, but to the economy and culture. This strategy would be typical of the second half of the 1990s–to talk about Russia where a neutral or even open position could be taken.

Some change came in 2001, when PiS's program included a passage heralding the narrative that was advanced by the party in subsequent years: "The pursuit of good-neighborly relations with the Russian Federation must be conducted with an awareness of the dangers associated, in particular, with the fact that our partner has numerous assets inside our country" (PiS, 2001). It is no coincidence that this occurred during the period when Vladimir Putin took power in that country. The Russian Federation became a threat to Poland in the military, political and economic fields, with a particular focus on energy. In some cases, the threat has been securitized. Lech Kaczyński, taking the oath of office before the National Assembly, said:

> An important issue is our relations with Russia, which has remained for centuries, despite the vicissitudes of fate, our great neighbour. We look at them taking into account, first of all, the historical perspective, maintaining patience and the conviction that there are no objective reasons why they could not be good. (Kaczyński, 2005)

"We would like to have the best possible relations with our neighbors in the East," Jarosław Kaczyński declared a year and a half later when taking over the

Prime Minister's portfolio (2006). – "There we must be, first, patient, second, patient and third, patient. The processes that will lead to Poland's acceptance as an important European actor and partner for Russia will probably be long" (Kaczyński, 2006). In doing so, he pointed out the boundary conditions for good relations between the countries – the superpower must recognize that Poland, as a member of the EU and NATO, holds a different status than when it left the Warsaw Pact and RWPG. These relations must be rebuilt from an equal position to Russia.

When winning the 2007 elections and taking over the Prime Minister's portfolio, Donald Tusk said: "while we have our views on the situation in Russia, we want dialogue with Russia as it is. Lack of dialogue serves neither Poland nor Russia" (Tusk, 2007). This approach was also characteristic of Platforma Obywatelska in the following years of the 21st century. After 2010, when PiS accused the Russian Federation and V. Putin personally of causing the Smolensk catastrophe, the PO stayed with its line, at least until 2014. Hence Bronisław Komorowski (2010), standing before the National Assembly declared: "There will be no stable development of our region without cooperation with Russia." A year later, in the parliamentary campaign, the PO wrote in the same vein in its program: "A constructive approach to Russia strengthens our position in Europe. (...) It is in our interest to dialogue with Russia primarily through the Union" (PO, 2011).

Meanwhile, the PiS program from that campaign was full of threads that not only described Russia as a threat but also pointed out the behavior of politicians in Poland who do not accept this stance as dangerous. It clearly indicated that failing to take a hard line against Russia amounts to treason. This narrative, developed across successive elections, bore the hallmarks of securitization: a new (or rather, old but with a renewed force of destruction) threat emerged, which had to be confronted by all means in various areas (political, military, economic, cultural). In the party's 253-page program, almost 10 pages are devoted to this diagnosis of Russia as an existential threat: "Russia, regardless of who governs it, will have no interest in strengthening Poland's position through the development of bilateral relations until it acknowledges that our country's position is robust enough to make maintaining good relations with it beneficial" (PiS, 2011).

The excerpt quoted above from the PiS program was absolutely unique compared to the material presented so far. Even during periods of particular tension between Poland and Russia, even among small, radical groups, there has never been such an openly hostile tone towards another country. This breakneck narrative frame: the threat from Russia – the EU, especially Germany, pressuring Poland – Poland dependent on Russia in the hands of EU powers, persisted in the language of PiS for a long time. The act of securitization was based on this frame. There was a DANGER (ex-KGB-ruled Russia, which was at least complicit in the death of the president), for which the only SOLUTION was a change in the position of prime minister (traitor, promoter of foreign interests), to which the grouping appealed to the voters (appeal for ACCEPTANCE through voting). Throwing accusations of a large caliber, believing in the evil intentions not only of the neighboring state but also of political opponents, and openly proclaiming this in a serious document while pointing out the multidimensional consequences of maintaining the threat, were intended to make the threat existential.

J. Kaczyński was assisted in maintaining the narrative of a special Russian threat by external circumstances, particularly following the seizure of Crimea in 2014 and the aggression in 2022. However, prior to these events, PO won the elections again, and Prime Minister D. Tusk's only remarks about Russia were that gas contracts with Russia, along with agreements with Qatar and the gas port, provided Poland with "gas security." His successor, Ewa Kopacz (2014), mentioned the consequences of deteriorating relations with Russia, such as embargoes on Polish food products, but did not address the underlying reasons. It was not until a year later, in the PO election program, that it took a stance regarding the new international situation:

> At present, Russia is damaging the chances for good relations with Poland by questioning the international order in Europe, undermining the principles of inviolability of borders, non-violence in international disputes and the sovereignty of states. We will consistently oppose Russia's aggressive policies and continue to successfully seek unity in the Western world on this issue. We believe that in relations with Russia, Polish interests are best served through the European Union on economic issues and NATO on security issues. (PO, 2015)

PO's position to talk to Russia through the EU or NATO has been a constant in the party's narrative since then. The party called for solving this problem (threat) through the rules/principles of normal politics, not extraordinary measures.

PiS, on the other hand, began its presentation of the Russian issue with a description of the threat:

> Russian imperialism is being reborn before our eyes. Russia is demanding a new balance of power in the world. It is breaking up the unity of our region and the Union and NATO. (…) The European Union is torn between wanting to cooperate with resource-rich Russia and wanting to

> maintain relations with post-Soviet countries that would like to be closer to Europe. NATO is also divided on this issue. (PiS, 2015)

The diagnosis led to the conundrum that Poland cannot be passive: "We cannot delude ourselves that Moscow will reward us for our passivity or tacit approval of Russian imperialism" (PiS, 2015). In this context, the diplomatic provisions of the PSL program looked as if they had been carried over from another era: "Strengthening good-neighborly relations on our eastern border, with Russia, Ukraine and Belarus, while supporting democratic processes in these countries" (PSL, 2015).

The situation was repeated after the next elections. In the campaign, PiS spoke of the clientelism of its political opponents toward Russia, a submissiveness that was supposed to ensure Poland's praise in the "salons of the European Union." The party then declared: "We want to improve relations with Russia. These are necessary not only for both countries and peoples, but also for the region and the continent. However, for changes to take place they require the will of both countries" (PiS, 2019).

PSL viewed Russia solely in economic terms – as a source for food exports, which are currently blocked, and coal imports, which should be halted. However, the program made no reference to the security issues at hand. The SLD and Konfederacja did not tackle the Russian problem. PO criticised the PiS government for its record coal imports from Russia, but did not elaborate on this point. Prime Minister Morawiecki, addressing parliament for the second time with a motion for a vote of confidence, emphasised the need for energy independence from the East, yet failed to address at least the allegations regarding coal imports. He did not portray Russia as a threat, nor did President Andrzej Duda in 2020. The situation changed dramatically during the 2023 campaign following the Russian aggression against Ukraine. However, an act of securitization did not occur, as there was no call for extraordinary measures. The national consensus on the necessity to invest in defense, build energy independence, and support Ukraine's struggle against Russia meant that decisions in this area were made according to a "normal policy."

The situation was different in 2023. "I think we all know what would happen if Russia triumphed in this conflict. I don't think we can pretend in this situation that we don't know the height of the stakes," Donald Tusk (2023) said in his expose without elaborating on the thread in a particular way. "Russia's neo-imperialist policy is only gaining momentum, we can see it in Ukraine," Mateusz Morawiecki (2023) conceded.

The narrative surrounding Russia in Polish political discourse has been shaped by three factors: (1) the collapse of the bipolar world and the dissolution of old alliances, (2) the changing leadership in Russia (from B. Yeltsin's "reeling" empire to V. Putin's rebuilding power), and (3) the attitude of the Western world toward Russia. Until the end of the 20th century, while Poland awaited the withdrawal of occupying troops from its territory and the opening of the door to NATO, the strategy of "not irritating the bear" prevailed. The clear signal from Washington and European capitals that Yeltsin had to agree to NATO enlargement meant that the narrative was extremely diplomatic: they preferred to discuss trade and culture rather than tanks or guns. The situation changed in the 21st century. V. Putin's rise to power coincided with a growing belief among the new countries of the North Atlantic Alliance that NATO membership guaranteed that Russia would not view Poland differently from Germany or France. However, Putin's strategy of negotiating with states rather than with NATO or the EU, along with the low effectiveness of these organizations in presenting Poland's perspective in relations with Russia, meant that at least from 2004/5 (Nord Stream 1) through 2008 (Georgia) to the crucial year of 2014 (the seizure of Crimea) and finally 2022, Russia began to be openly portrayed by some political circles as an enemy (an existential threat) and by others at least as a problem that needed to be handled with care. Although the strategies for addressing this threat varied (bilateral, strong alliance with the US with military presence on Polish territory, or allied actions), due to the change in the attitudes of leaders from other EU and NATO countries after 2014, Russia began to be treated as a "security issue." This became part of the securitization process – a justification for changing defense strategy and increasing military spending. The securitization of Polish-Russian relations and Russian imperial policy is presented in Figure 13.

FIGURE 13 Securitization of the relations with Russia
SOURCE: OWN STUDY

3 Discourse on Terrorism

Terrorism as a threat to both global and local security did not appear in electoral programs and political speeches until after the September 11, 2001 attacks, as Leszek Miller (2001) said in his exposé delivered just over a month after the attacks on New York and Washington: "The tragic events dramatically reminded everyone of the old truth that individual and overall security is not given once and for all" (Miller, 2001). Two years later, when Poland took part in the operation in Iraq, the terrorist threat was a side thread justifying the involvement. Announcing the decision, President Aleksander Kwaśniewski said that "we were intervening in the belief that together we should take action against terrorism" (Kwaśniewski 2003). "Poland must also look at terrorism and the threat of weapons of mass destruction as potential threats to our country as well," Leszek Miller (2003) said a week later during a debate on involvement in Iraq.

Despite the above, terrorism did not take up much space on political agendas in subsequent years. In the 2005 campaign, only PiS and PO referred to it, in general way: "International terrorism remains an immediate threat" (PiS, 2005). "Terrorism, organized crime, uncontrolled migration are much more serious threats today" (PO, 2005). "The current challenges of world politics, such as globalization and terrorism, impose a close alliance between the European Union and the US," Prime Minister Kazimierz Marcinkiewicz stated briefly in his exposé (2005). Slightly more space was devoted to this threat a year later by Jarosław Kaczyński (2006): "Today this external security revolves primarily around two problems, that is, economic security and security against terrorism" (Kaczyński, 2006). In the next election, the Law and Justice Party proposed stricter penalties for terrorist crimes.

"Global policy does not revolve in a vacuum. Still, it centers around specific problems and challenges: energy security, climate change, terrorism, migration, hunger and disease, debt, weak and failed states" (LiD, 2007), the Left's program said, placing terrorism in line with other global threats. The PO spoke in a similar vein, noting that "only a strong and efficient EU can meet the challenges that no member country will be able to face alone: terrorist threats, migration pressures, climate change" (PO, 2007). Later in the program, PO noted that "Poland could become a transit state for terrorists at any time. It is also impossible to rule out terrorist acts and attacks in any of Poland's cities" (PO, 2007).

Terrorism, on a global scale, has posed a threat that has been subject to strong, effective, and – as Rita Floyd advocates – in some cases, like the war on terror as a response to this threat, unjust – securitization. Spectacular attacks that can be executed at minimal cost anywhere in the world have led most societies to accept extraordinary measures for combating this threat: from banning water bottles on aeroplanes to the humiliating removal of shoes during personal inspections, alongside special laws permitting services to wiretap, arrest, and imprison suspected terrorists, both domestic and foreign. Poland, however, did not embrace the securitization shift of 2001. Although the attacks on the USA shocked Polish citizens and policymakers, there was a prevalent perception that these events were a geographically distant problem. This situation only changed at the beginning of 2015, when Europe started to experience attacks carried out on its territory by jihadists (Madrid, Nice, Manchester, London, Paris, Brussels, Berlin), in which Polish victims were also involved. The large mass events held in Poland at that time (NATO Summit, Youth Days with the personal participation of Pope Francis) and the associated threats necessitated new legal solutions, such as the Anti-Terrorism Act of 2016. However, beyond this, we cannot speak of an act of securitization here.

References to terrorism in the 2011 elections were more specific than in previous years and included fairly detailed arguments. The PSL called for the protection of infrastructure. PiS indicated that "the preparation and enactment of a special law on countering terrorist threats was becoming an increasingly urgent need" (PiS, 2011). The Left did not believe in the state's readiness to fight terrorism. 10 years after the 9/11 attacks, terrorism was present in election programs, but the scale of this presence was not large. The topic only returned in the party's 2015 PO election program, as a demand to strengthen the anti-terrorism system.

> Despite the persistently low level of terrorist threat in our country, measures are needed to expand the scope of cooperation of institutions that care about anti-terrorist security, to make citizens aware of the possibility of events and the rules of conduct in the event of their occurrence. (PO, 2015)

The issue was presented somewhat differently by PiS, which in its program advocated changes in the law and new powers for individual services, announced as early as 2011. PiS was given the opportunity to implement its demands. Upon taking the prime minister's seat, Beata Szydło (2015) began her exposé with the terrorist threat precisely, referring to the Paris attacks that took place a few days before her speech: "I take office as Prime Minister of the Republic at a time when our world – built on the values of freedom, democracy and tolerance – is increasingly aggressively attacked by those who do not respect the right to freedom of others. We do not agree with such a vision of the world" (Szydło, 2015). References to the terrorist threat in Prime Minister Szydło's speech were not without pointing out that one of the causes of attacks in Europe is refugees and the flawed migration policies of Western member states.

In 2019, terrorism was no longer a significant problem. In its program, PiS stated: "We have no record of acts of terrorism in our country, and there are no radical Islamic violent organizations operating in Poland" (PiS, 2019). KO announced the repeal of some provisions of the anti-terrorism law that harm civil rights: "We will protect the privacy and secrecy of correspondence of citizens and citizens. Those provisions of the 'surveillance' and 'anti-terrorism' laws that may violate human rights and civil liberties will be repealed. They were adopted in defiance of a Constitutional Court ruling" (KO, 2019). Prime Minister Mateusz Morawiecki (who did not mention terrorism once in 2015) simply stated in his exposé that *new threats*, "such as global terrorism and waves of immigration, have emerged in the world in the 21st century" (Morawiecki, 2019). Nor was terrorism a topic of the 2023 campaign.

4 Discourse on Crime

An important part of political agendas and speeches by politicians is occupied by issues related to law and order within the traditional approach to security. Today, these issues are central to homeland (internal) security analysis, and their social significance ensures that the topic's prominence in political discourse remains strong.

In Polish political discourse, one can distinguish three periods with varying distributions of accents and narratives in the area. The 1990s was a time for shaping the system to ensure internal security: the construction of both the judiciary and the law enforcement apparatus (the reform of the police and the establishment of specialized structures to combat crime) took place under the challenging conditions of rising organized crime. This period saw the creation of a new criminal code (1997) abolishing the death penalty while advocating for increased penalties for crimes, laying the groundwork for anti-corruption legislation, and highlighting the chronic underfunding of services. A significant symbol of this era is the assassination of Police Chief Gen. Marek Papała for motives and under circumstances that remain not fully explained. Symbolically, the establishment of a professional formation to combat organized crime – the Central Bureau of Investigation (CBŚ, today: Police Central Bureau of Investigation – CBŚP) – in April 2000 can be considered its conclusion.

The years 2000–2007 were a time when these issues became particularly politically important. Lech Kaczyński built a political career on his fierce opposition to the liberalization of punishments and the sluggishness of law enforcement agencies, earning him the nickname "Sheriff": first as Minister of Justice in Jerzy Buzek's government, then as Mayor of Warsaw, and finally as President of Poland. Symbolically, this period ends with the creation of the Central Anti-Corruption Bureau (CBA) – a new special service established to combat corruption and actions detrimental to the state treasury.

The second decade of the 21st century is a time when issues of law and order have given way to problems such as terrorism, migration, and external threats to the entire Republic. In the political discourse, there are themes of service oversight, financing formation, and judicial reform.

In the 1990s, a key issue was the organization of the law enforcement and justice apparatus, along with reforms of the criminal code and laws inherited from the PRL. Sovereignty, independence, rule of law, and efficiency/effectiveness emerged alongside the concepts of justice and security. Potential threats to the rights and lives of citizens were perceived not only from criminals but also from the state and its institutions, as well as from the remaining post-PRL legal regulations. Calls for ensuring a just verdict signified not only a demand for punishment for the offender, but also for the protection of the citizen from wrongful conviction, along with guaranteeing him a fair trial and respect for his rights. Judicial reform was one of the key issues raised in the exposé by Prime Minister Tadeusz Mazowiecki. "We will strive to accelerate our work on new codifications of criminal law, particularly criminal procedure, which must meet the requirements of the right to a defense and the right to a court," said the former oppositionist and Solidarity activist, emphasizing not fighting crime, but ensuring human rights (Mazowiecki 1989) In a similar vein, he characterized the tasks of the transformed militia: it was to become a formation that supports rather than harasses citizens. Jan Krzysztof Bielecki spoke about the withdrawal of state supervision of citizens in every sphere of their lives, but also very clearly stated that: "criminal crime is becoming an increasingly dangerous scourge. The development of democratic freedoms cannot be compromised by the impunity of bandits and thieves" (Bielecki, 1991). He pointed to not only low salaries in the police, but also the lack of decisiveness caused by its transformation, personnel movements, and uncertainty about the expectations of superiors as reasons for this state of affairs. Waldemar Pawlak (1992), analyzing privatization processes, spoke of the threat of corruption:

"Very dangerous to the reform process are the phenomena of corruption, fraud and getting rich by seizing public pennies. We will fight these phenomena decisively."

The 1993 campaign already focused more on citizens who were afraid of crime.

> "We must begin to feel safe in our homes and on the streets. Rising crime must be decisively opposed by the state bodies set up for this purpose. We consider it necessary to tighten penal policy," PC declared. – "When it comes to punishing crimes, it is necessary to introduce the possibility for courts to impose prison sentences combined with hard labor. The rights and interests of crime victims must be respected in criminal proceedings." (PC, 1993)

This would be a narrative characteristic of this party (later PiS) for the next 20 years. Other right-wing and conservative groups also spoke in a similar vein.

The left and the center parties generally did not deviate in their narrative from right-wing groups, but focused more on the justice system's effectiveness. SLD called for creating "an Efficient System of Legal Protection Institutions to guarantee that criminals would receive just punishment" (SLD 1993), while UD wrote: "A growing problem is the low efficiency of the justice system. This gives rise to insecurity and loss of citizens' sense of security" (UD, 1993).

Prime Minister Pawlak's second exposé (1993) already presented crime as a serious problem.

> "People should not live in fear of losing their lives, health, property. Efforts should be made to remove the causes and reduce organized and economic crime, to fight corruption, bank fraud, tax fraud, money laundering, car theft, drug production and trafficking. There will be a rapid strengthening of crime-fighting structures," Pawlak declared. (1993)

His successor, Józef Oleksy, continued: "Inevitable and swift punishment must be one of the basic elements of prevention. After all, it is the criminal who is supposed to feel threatened, not the citizen" (Oleksy, 1995). W. Cimoszewicz did not break from the narrative of his predecessors. He portrayed crime as the biggest scourge facing the state.

> The growing number of crimes committed and the dangerous tendencies involving a rapid increase in the number of crimes of the most serious types, the emergence of organized crime, and finally the criminal world's

> building of a strong financial base for its activities – pose a challenge to the state. (Cimoszewicz, 1996)

The problem of crime also found a place in the president's speech. In 1995, during his swearing-in, Aleksander Kwaśniewski said that "we cannot be the first generation to live with a sense of external stability, and at the same time the first to have a sense of instability and fear on the streets of cities." The parliamentary election campaign of 1997 was marked by increasing efficiency, tightening up, improving functioning and order. Unia Pracy, in the very first words of its program, promised "a determined fight against crime and corruption." The ROP declared: "We will curb banditry and crime. We will streamline the judiciary, reducing the length of court procedures to the maximum" (ROP, 1997). AWS promised changes in the law to make it easier to punish criminals. Other parties spoke in a similar vein. SLD declared an "effective fight against juvenile crime and demoralization" (SLD, 1997), PSL promised to "protect the life, health, property of citizens and ensure public order" (PSL, 1997). The UPR wrote in its program: "Citizen security means efficient police, prosecutors and courts. The inevitability and permanence of the punishments handed down" (UPR, 1997).

> "First – free people in strong families. Second – repair of the state," the new Prime Minister Jerzy Buzek (1997) presented his government's action plan. – "Citizens must feel safer than they do today. This requires improving the work of the police, other services, law enforcement agencies and courts. The justice system is the foundation on which a strong state must be based. Meanwhile, it is in deep crisis." (Buzek, 1997)

Three years later, Aleksander Kwaśniewski spoke in a similar vein as he began his second presidential term: "I also want another issue of importance to the world of politics is the duty to raise the level of citizens' sense of justice and personal security. This is about the effectiveness of the courts, prosecutors, police. It's about the general disapproval of corruption" (Kwaśniewski, 2000).

The 2001 parliamentary campaign was full of references to crime and internal security. The reason for this was the seriousness of the problem. As the SLD noted in its program, "although successive governments announce a fight against crime, the situation is not improving. People still feel threatened. Victims and victims of crime wait too long for the perpetrators to be detected and tried" (SLD, 2001). Samoobrona and PSL spoke briefly about the need to improve the fight against crime: "Eliminating the causes of the increase in crime, not just the effects, confiscating the property of affaires and fighting corruption" (Samoobrona, 2001); "Increasing the effectiveness of the state in maintaining order, order and security for citizens" (PSL, 2001). "In Poland, there is a growing sense of helplessness in the face of crime, leading to a widespread decline in security that citizens feel. The basic problem is the state's lack of sufficient capacity in introducing a unified crime-fighting policy and in consistently implementing it." – seconded by PO (2001).

PiS devoted the most space to this topic. The party's 2001 campaign was based on a program to fight crime. It can even be assumed that in its first campaign, this party was a one-issue party, which allowed its leaders to build political capital in the following years. No other party in Poland after 1989 has presented such an approach. PiS appealed to the vision of criminalizing the economy, portraying it through the prism of corruption and fraud, questioning how the transition to a free market in the 1990s occurred and building the belief that certain socio-political elites profited from it. Hence, PiS called for a referendum with a question on introducing a law to tighten the fight against crime. The program went on to enumerate in detail: "It is necessary to carry out a fundamental revision of the Penal Code, so that its basic principles of punishment and the clear increases in the threat for individual crimes are changed" (PiS, 2001). The party's proposals included much more far-reaching solutions than those of other parties. The new special laws, the extraordinary mode of punishing the most dangerous offenders, and the search for public support through a referendum demonstrate the securitization of the issue of fighting crime that has been undertaken – moving it from ordinary politics to the category of an emergency (existential threat), requiring exceptional and extraordinary measures.

PiS made heavy accusations by encasing them in a typically populist narrative: only the rich elite are doing well, and only they can feel safe. "An average citizen who makes a mistake in his tax return, a small businessman who fails to pay his tax on time are treated as criminals. Meanwhile, the real criminals, who have made fortunes, sleep soundly" (PiS, 2001). Hence, there is a demand for the services to be given greater powers to obtain information potentially usable in fighting crime. As mentioned, all parties pointed to the need for change, but none presented the entire sphere as pathologized to the marrow and in need of immediate, far-reaching "corrective" measures. PiS, however, needed it to carry out the securitization move smoothly (Figure 14).

Despite a good result in the poll, PiS did not receive enough support for its proposals. The elections were won by the Left and it was SLD, together with the PSL, that formed the new government. On taking office as prime minister, Leszek Miller said Poland must become a safe country. Moreover, he declared: "We declare an unwavering fight against criminals, regardless of their identity.

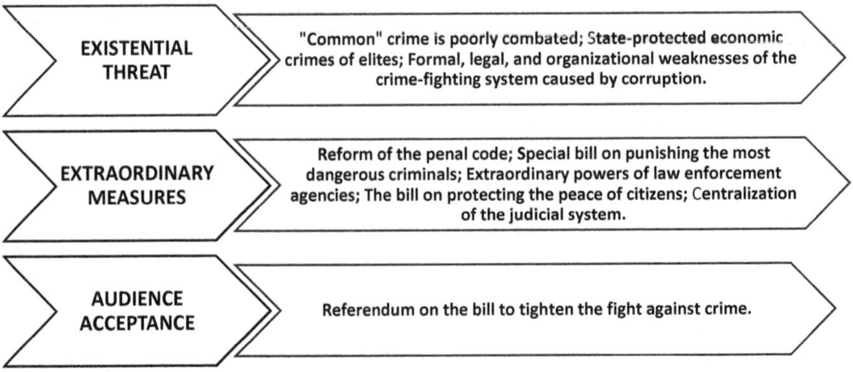

FIGURE 14 Securitization of the crime in the political discourse of PiS in 2001
SOURCE: OWN STUDY

(…) Justice must proceed swiftly, as it is then more effective and economical" (Miller, 2001). This was a narrative held in the same vein as the campaign proposals, not as radical as that of PiS.

The three major corruption scandals uncovered under the SLD and PSL governments instilled in some of the public a belief that the claims of a corruption-ridden PiS elite might have been accurate. During the 2005 campaign, all parties proposed changes in this regard, but PO and PiS had the most to say, although the tone of these parties was quite different.

PiS's program content can be divided into three aspects. The first focuses on holding predecessors accountable, the second addresses legal changes to combat crime, and the third deals with reforming institutions and services. The program began by announcing the establishment of "a parliamentary Truth and Justice Commission. Its primary task will be to investigate all types of scandals and abuses" (PiS, 2005). The proposal for extraordinary solutions was already evident in this first provision. The state of security was assessed in the context of the story of the bad wolf.

> Murders of random passersby in housing estates, robberies and rapes of defenseless women, thefts of cell phones, car burglaries, and numerous economic scandals confront Poles daily. As a result of liberal penal policies and the poor functioning of the police, courts, and prosecutors' offices, the number of crimes remains very high. Instead of sitting in jail, bandits roam the streets of cities, feeling a sense of impunity. (PiS, 2005)

The story pointed directly to the guilty and promised to hold them accountable. When the frightened reader began to check behind his or her back to see if a threat was lurking, a "deliverance" appeared – a program of recovery. It was detailed in over a dozen pages and addressed the smallest details of the issues raised. "The new Criminal Code will effectively combat crime and protect victims of crime. The high risk of severe punishment will effectively deter both petty thieves and dangerous bandits" (PiS, 2005).

The issue of punishment received considerable attention in the program. It was central to both combating and preventing crime. Other measures – beyond the deterrent effect of a high sentence – were not envisioned by Pis. The proposed range of punishments was extensive and varied, including longer prison sentences and the introduction of absolute life imprisonment, higher penalties for repeat offenders, confiscation of criminal property, and tougher penalties for violent crimes. In return, citizens who abide by the law were to be supported in protecting their rights. After announcing the aforementioned changes in the law, PiS turned to organizational issues and highlighted elements of reform for the judiciary and law enforcement apparatus. Just as it did not believe in the possibility of prevention beyond deterrent punishment, it also assumed that there was a pathology in the courts or the prosecutor's office that needed to be eradicated. This pessimistic worldview fostered a sense of insecurity, thereby laying the groundwork for the securitization act. "Significant organizational changes are essential in the prosecutor's office. We will establish an elite unit within the central prosecutor's office to combat major mafia and scandal-related crimes" (PiS, 2005).

As in the 2001 campaign, the issue of fighting crime was securitized. We have a constructed threat and a proposal for extraordinary measures (new powers for the services, new obligations for businesses). Acceptance of the extraordinary measures was intended to be the vote cast in the elections (Figure 15).

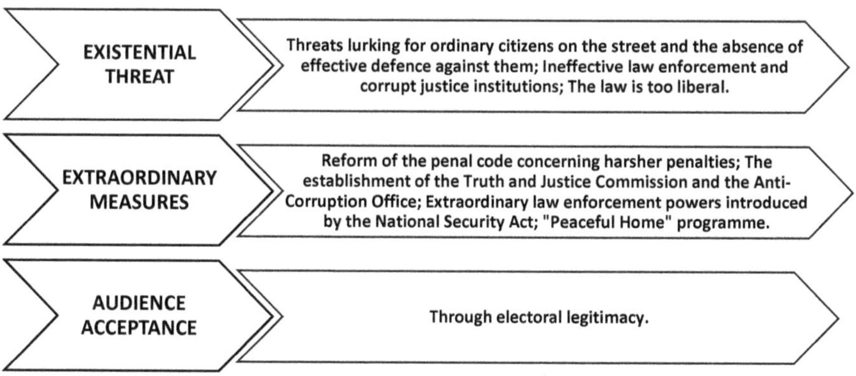

FIGURE 15 Securitization of the crime in the political discourse of PiS in 2005
SOURCE: OWN STUDY

The PO in its program also criticized the state of security and the functioning of the institutions providing it, though not entirely. Although critical, the diagnosis of the state of security was conveyed in a language distinct from that of PiS. The PO addressed the same problems pointed out by PiS and made similar accusations. However, the program included objections regarding the limits of deliberation of codes and the empowerment of services. There was no call for extraordinary measures, and the issue of crime as a threat was not raised to an existential level. Thus, there was no securitizing procedure here; the problems remained within the framework of ordinary politics.

Voters, however, opted for harsher rhetoric and allowed the Law and Justice party to win the dual elections – parliamentary and presidential. Kazimierz Marcinkiewicz, who took over as prime minister, continued his party's narrative:

> This is not the kind of state we dreamed of when we regained sovereignty in 1989. Corruption, in addition to public outrage, spoils market mechanisms. The low efficiency of law enforcement and the judiciary reduces the security of economic transactions. It creates a sense of impunity in criminals and vulnerability and helplessness in citizens. (Marcinkiewicz, 2005)

Later in the exposé he declared: "we want to take back the state from the pathological arrangements and informal pressure groups parasitizing it. We want to give it back to citizens" (Marcinkiewicz, 2005). Replacing Marcinkiewicz as prime minister, Jarosław Kaczyński upheld the line outlined by the party: "We need to go further here. We have to make the Pole know that the right to peace in the apartment, on the street, at work is his right guarded by the state" (Kaczyński, 2006). In the 2007 campaign, PiS reiterated most of its earlier demands. It slightly reframed the narrative due to its two years in power: there was no longer a sharp criticism of the status quo, though it was made clear that the reforms had been interrupted by early elections. Given previous assessments of how much remains to be addressed, one might question the program's final conclusion. Nonetheless, the fight against crime played an important role, as it had in 2001 and 2005, although it no longer exhibited the hallmarks of a securitization act: the anti-crime plan was incorporated into the political agenda and began to be governed by the rules of standard politics.

PiS's biggest rival, PO, also referred to the issue of fighting crime and organizational changes in this area. Still, the perspective adopted in the narrative focused on correcting the mistakes of predecessors: "Our goal is to reduce threats to state security and public order while increasing the citizens' sense of security, without unduly restricting their freedoms and liberties" (PO, 2007). As in earlier programs, PO proposed tightening regulations or granting new powers to the services, but clearly stipulated that the process must not violate civil liberties. Partia Pracy spoke similarly: "Restoring confidence in the justice system. Addressing real crime, including violence against women and children and trafficking in women and children" (PP, 2007). Donald Tusk's (PO) addressed the issue of fighting crime at the very end: "The fight against crime will be carried out with full determination. But it is not enough to introduce only toughening punishments and dazzle the public with spectacular arrests" (Tusk, 2007). In an attempt to regain power in 2011, the Law and Justice party returned to "zero tolerance" rhetoric and criticized opponents for being too liberal in their approach to fighting crime. It recalled its successes in this field, interrupted by the loss of early elections, and promised to continue the program announced previously. The diagnosis made by the SLD was similar to that made by PiS. The party accused its predecessors of not doing enough and of manipulating statistics.

> The sense of security in our country is still far from expectations. Most of us fear traffic pirates and drunk drivers, theft and burglary, hooligan attacks and assaults. We are not sure that our children are effectively protected from drug traffickers, as well as from various forms of violence and harassment. (SLD, 2011)

The parties in power for the previous four years did not question the reality but only planned further actions. Donald Tusk indirectly referred to the internal security issue, talking about reform the pension system for law enforcement. Three years later, Ewa Kopacz asked rhetorically: "How do we find the line between what is necessary for people to feel safe and their natural need for freedom?" (Kopacz, 2014).

"The competency chaos in homeland security and crisis management will be eliminated first" (PiS, 2015) – declared PiS in victorious 2015 campaign. The Left hinted at the need to control the work of the secret services. PO presented the set of slogans without details (PO, 2015). Prime Minister Beata Szydło, speaking on behalf of the victorious PiS, said in her exposé: "We will accelerate the modernization of the uniformed services, restore proper operation and supervision of the special services" (Szydło, 2015). However, no details were given. It was the last time, when law and order issues were raised in the campaigns and prime ministers' speeches. The time when political capital could be built on these topics has passed.

5 Discourse on Corruption

As with crime in general, certain phases can be distinguished in the narrative and political discourse on corruption and organized crime. The first is (1) the 1989–2000 period, when corruption was stigmatized, but solutions to combat and prevent it were lacking, as in the case of economic crime. The second – (2) the period 2001–2006, namely the time of building laws and institutions, including the creation of the Central Anti-Corruption Bureau (CBA). The third period – (3) from 2007 to the present features a variety of proposals for reforming the system and improving the effectiveness of prosecution.

In the 1990s, politicians spoke about corruption with one voice, as to the need to combat it. Prime Ministers T. Mazowiecki, J. K. Bielecki and J. Olszewski did not yet consider it a topic worthy of an exposé, but Waldemar Pawlak was already declaring firmly: "The phenomena of corruption, fraud and getting rich by seizing public pennies are hazardous to the reform process" (Pawlak, 1992). His successor, Hanna Suchocka, was even more resolute: "I will immediately demand that ministers review the criteria and procedure for granting concessions and quotas. The government will introduce transparent criteria and an open procedure in such matters" (Suchocka, 1992). "Efforts should be made to remove the causes and reduce organized and economic crime, to fight corruption, bank fraud, tax evasion, money laundering," W. Pawlak declared in his next exposé (1993), when he headed the SLD-PSL government. Józef Oleksy and Włodzimierz Cimoszewicz did not raise the topic, as did most party programs from the 1997 elections. The issue of corruption or economic crime appeared rarely.

During his swearing-in for a second term, President Aleksander Kwaśniewski said, listing Poland's successes: "Many problems have not been solved. I'm thinking of the scale of (...) corruption" (Kwaśniewski 2000). The turn of the century was the beginning of treating the phenomena of corruption and economic crime as serious challenges that needed to be solved, if only in the context of Poland's efforts to join the European Union. In the 2001 campaign, more space was already devoted to these issues. There appeared for the first time a demand – put forward by PiS – for the establishment of a central anti-corruption office: "new formation with strong powers, made up of new people, not entangled in the arrangements, which could take action to identify and expose links between politicians and the criminal world" (PIS, 2001). The Left pointed out that "a more effective fight against corruption was an extremely urgent need" (SLD, 2001), and Samoobrona proposed: "Confiscation of property for scandals and the fight against corruption, accounting for privatization: who privatized what, for how much and where the money is" (Samoobrona, 2001).

Corruption was included in the exposés of both prime ministers of the SLD-PSL government.

> "Mess, waste of public good, conflicts, incompetence and ineptitude were features of the past term. It was challenging to find a month without scandal, suspicions of corruption, examples of privatization and power quarrels," Leszek Miller began his speech. – "Criminal repression of corruption crimes should be tightened, including confiscation of illegally acquired property." (Miller 2001)

Marek Belka also spoke similarly: "It is worth talking less about corruption and acting more against corruption" (Belka, 2004).

In the 2005 campaign, the subject was raised even more frequently, and the proposal to establish a special office to combat corruption was raised not only by PiS. The issue was also one of the main topics of a speech by Kazimierz Marcinkiewicz, who represented PiS as prime minister: "Corruption, in addition to public outrage, spoils market mechanisms. The low efficiency of law enforcement and the judiciary reduces the security of economic turnover" (Marcinkiewicz, 2005). The narrative was continued a year later by Jarosław Kaczyński when he took over as prime minister: "This is a pathology, at the center of which is corruption as this phenomenon that is destroying our social life. This is a pathology related to the use of public funds for private purposes. This is the building of mafia arrangements around the state apparatus" (Kaczyński, 2006).

PiS's proposal to establish an anti-corruption office and start a "new" IV Republic can be seen as an extraordinary solution to combating a significant threat. Never before – or since – has there been a proposal to establish a new, separate special service. To date, the service has only been reformed. State anti-corruption security is another securitized area in the political discourse of the Third Republic. Fig. 16 illustrates the process vividly.

Corruption and fight against it remained one of the main elements of PiS's anti-crime narrative. In the campaign before the 2007 elections, the party's program meticulously enumerated the successes achieved in this field: "As promised, we established the Central Anti-Corruption Bureau. The CBA fights corruption regardless of political options, views, wealth of wallet" (PiS, 2007). PO also addressed the issue of fighting corruption, but was not so radical: "To fight corruption, it is essential to adopt a cross-party anti-corruption pact, defining methods to fight this phenomenon. It must include not only organizational, detection and procedural measures, but also preventive ones" (PO, 2007). PiS was not able to continue reforms in this area. The incoming

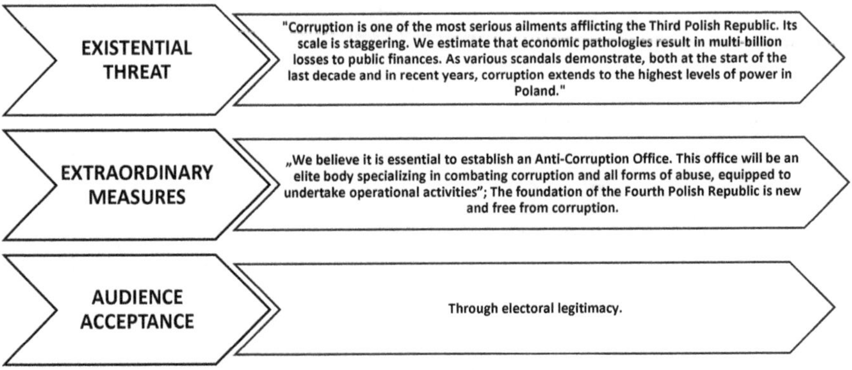

FIGURE 16 Securitization of the corruption in the political discourse of PiS
SOURCE: OWN STUDY

Prime Minister Donald Tusk (PO) understood that fighting corruption was still a hot topic and devoted a lot of space to it in his exposé. To show that he also has initiatives in this area, he announced the appointment of an anti-corruption attorney (Tusk, 2007). The subject was also revisited in the 2011 campaign, where PiS recalled its successes before the PO-PSL coalition government. SLD was as critical as PiS, although it looked elsewhere for the culprits and defined the causes of the phenomenon differently. It pointed out that "one of the reasons for the occurrence of corruption in individual cases is the still existing deficit of services in certain public sectors" (SLD, 2011). However, in the case of crime, the era for gaining political capital in the fight against corruption was to be over. Donald Tusk's second exposé (2011) did not devote even one sentence to corruption. Three years later, Ewa Kopacz (2014) also did not consider the problem important enough to talk about it on the day of her speech to the Sejm. In the 2015 campaign, even PiS did not devote a separate space to the topic, and the PO stated laconically: "We will introduce further solutions to improve the effectiveness of preventing and combating economic crime" (PO, 2015).

In 2019, the PO did not address the topic at all, while PiS devoted a single paragraph to it, focusing not so much on corruption itself but on its new key term: "the VAT gap." The concept, however, did not become an important enough theme to build a broader narrative around it. The topics of corruption and the fight against corruption in conjunction with economic crime have been desecuritized – they have returned to the bosom of ordinary politics. In 2023, the campaign talked about corruption only in the context of the scandal with the sale of visas to Poland by Foreign Ministry officials: this corruption led to an uncontrolled flow of migrants to Poland (PO, 2023). However, the issue has disappeared from the main positions of Polish political discourse on security.

6 Conclusions

The discourse on political security in Poland after 1989 is characterized by various topics and attitudes, including the securitization of certain issues. Relations with international organizations crucial to Poland's security, such as the European Union and NATO, were not securitized, despite the many efforts and sacrifices that the country made during the accession process. This was the result of a cross-party consensus that integration with Western institutions was both inevitable and strategic in terms of ensuring national security. Therefore, securitization does not appear in the debate on integration matters, and the aforementioned efforts are undertaken as part of "normal politics."

Periodic securitization pertains to relations with two key state actors for Poland's security: the US and the Russian Federation. In the first case, securitization involves the establishment and maintenance of special relations, particularly ensuring the presence of US troops on Polish territory. Extraordinary measures include, among others, the purchase of US weapons, funding for Americans, and the participation of Polish troops in foreign operations initiated by the US (Iraq, Afghanistan). In the second case – Russia – the imperial, aggressive policy of this country towards others, particularly those in the former Eastern Bloc, is portrayed as an existential threat to Poland, necessitating a response not only through increased spending on its own army but also by strengthening alliances, especially with the US, regardless of the costs.

The issue of fighting crime and corruption is also securitized, but only by PiS, which was founded as a "single-issue party" dedicated to fighting crime and corruption. Hence, the narrative of an existential threat, a cancer eating away at the Polish state, must be addressed differently than before – in a non-standard way – with new laws, manual control of the courts and the prosecution, deep systemic reform, verification of judges and prosecutors, and even media pressure. However, this is merely the demand of one party, which finds selective acceptance in society, as reflected in the results of parliamentary and presidential elections.

CHAPTER 5

Economic, Societal and Environmental Security Discourse

The subject of the economy's condition and state finances has always been present in public discourse. However, it has gained a security dimension relatively recently. Energy security was the most popular topic, especially in the second decade of the 21st century. It was primarily understood as ensuring the availability of energy resources from diverse sources (strategic resources, the lack of which or uncertainty in access to which is an existential threat), along with the state of the transmission and processing infrastructure and the condition of the mining industry.

References to economic security also appeared in the context of the overall condition of the state's finances and the economy, where, as in energy, it was sometimes combined with calls for the need for state interventionism, central planning, or nationalization (of banks, media). Part of economic security also included food security, particularly in relation to the state of agriculture. In the area of economic security, the issue of public infrastructure (especially roads) also emerged in Polish political discourse.

Societal security, in turn, presents four dimensions in Polish political discourse. The first is social security. Here, we are dealing with demands for developing a welfare state, reforming the insurance system, or supporting the excluded. The second dimension concerns health security, while the third dimension focuses on the issue of aliens – both potential immigrants to Poland (Europe) and, on a broader level, relations with the Polish diaspora scattered around the world, where the obligation to care for them and ensure the observance of their rights in the countries they inhabit is emphasized.

Regarding environmental security, it initially emerged in discourse solely within the realm of ecology, only to gain significance in the 21st century and be regarded comprehensively through the lens of its various dimensions.

1 Financial Security Discourse

The state of Poland's finances and economy after the 1989 breakthrough figured prominently in election programs and political speeches. At the time, the economy and the state's finances, or rather, threats to their well-being, were not securitized and therefore placed in the security category. If this occurred, it was only rarely, as the following examples illustrate. Even in these instances, however, the state of the economy is often linked more closely to the independence or sovereignty of the state, and thus only indirectly to security: "Poland must be a state of social justice, economically and politically independent" (Party X, 1991). Although there was a political definition of "state economic security, understood as an adequate state of production and extraction of energy resources, energy production, defense industry" (SLD 1991), but these were isolated examples of such activities. Straightforwardly, security and economy were first linked in 1991 by the Porozumienie Obywatelskie Centrum: "Poland's security and the well-being of its citizens depend on the state of the economy" (1991). "The basis of the country's security will be – increasingly – an effective economy aimed at a fuller utilization of Poland's personal and material potential," Stronnictwo Demokratyczne (SD, 1991) wrote in its program.

The answer to the question of why, until the end of the 1990s, an issue so vital to the survival of both the state and its citizens (and therefore existential), such as the transition of the economy from socialism to capitalism, the condition of the Polish currency – the zloty, household budgets, interest rates in banks, etc., was not securitized, is not obvious. It seems that, next to military security, this area should be second in the list of securitized issues, especially at a time when the fragile foundations of the economic order were easily compromised, unemployment was rising, the currency was devaluing, and society was finding it increasingly difficult to cope with the costs of the transition. Meanwhile, these issues were not labelled as security problems (thus the constitutive "speech act" did not take place), even when extraordinary measures were proposed and broad public acceptance was sought. The situation exemplifies how non-obvious and unpredictable the process of securitization is and how important the initiating "speech act" is. If a problem, even an existentially important one, is not named as a security threat, it will continue to function in the realm of politics until someone connects it to security.

When taking office as prime minister, J. Oleksy (1995) said that: "Entrepreneurs rightly expect security for their money in banks." His successor, W. Cimoszewicz, referred to the security of economic turnover and doing business in general: "The government is aware that the condition for maintaining positive trends in the economy is favorable rules and security of doing business" (Cimoszewicz, 1996). The economy was not presented as a subject of potential threats that could result in existential insecurity for the entire state, there was no mention of its condition, and only attention was paid to certain aspects that make it difficult to attract investment. The demand for "increasing the security of economic turnover" was included in the SLD's 1997 election program. Unia

Pracy diagnosed: "Economic security will be fostered by greater diversification of the directions of economic cooperation" (UP, 1997).

"National security has a dimension beyond the military. Equally important is economic and social stability," Prime Minister Jerzy Buzek (1997), representing AWS, said in his exposé, laying the foundation for a holistic view of the economy and its economic successes through the prism of security. However, it was not followed up: the 2001 campaign did not include references to economic security. Nor did the embracing year Alexander Kwaśniewski or Leszek Miller (2001) talk about it. "Dangerous phenomena growing in public finances" were mentioned in his exposé by Marek Belka (2004). A year later, Kazimierz Marcinkiewicz (2005) returned to the issue of "the security of economic turnover, lowered by the low efficiency of law enforcement agencies." Jarosław Kaczyński (2006) placed economic security (but understood narrowly, as energy security) as one of the key aspects of external security. "Today, external security revolves primarily around two problems, that is, economic security, particularly energy security, and security against terrorism" (Kaczyński, 2006). "Only a rapid increase in earnings will guarantee young people an adequate level of economic security, which is necessary to make a responsible decision to start a family and have children," PO wrote in its 2007 program, taking a personal security perspective.

As can be seen, for the first twenty years of the analyzed period, the connection between security and economy (economics) was made haphazardly, piecemeal, without being placed in a broader context. The word security (danger) served as an amplifier, the meaning of which was not analyzed. Economic (security) as a full-fledged category appeared in election programs and political discourse in the second decade of the 21st century.

The full subject scope of economic security – as it is understood – was presented by the PSL in 2011.

> In terms of ensuring the country's economic security, PSL has set itself the task of: accelerating GDP growth, rationalizing public spending, increasing the effectiveness of tax enforcement, introducing a system of tax breaks, consistently freeing the economy from bureaucratic constraints, increasing the role of entrepreneurs in creating laws that favor the creation of new jobs, supporting the activity and innovation of Polish entrepreneurs, developing trade contacts, continuing to support activity and innovation, strengthening the role of municipal governments' own revenues, creating land use plans in each municipality, including to eliminate haphazard investment locations, accelerating the modernization

> and expansion of power grids, expanding power generation based on domestic raw materials. (PSL, 2011)

Donald Tusk's entire second exposé of 2011 was devoted to overcoming the crisis.

> The fact that Poland has survived these four years in good shape, the fact that all around us this hurricane of financial crisis threatens the security of the most developed countries, the fact that all around us millions of people accustomed to tranquility, prosperity and security are today facing the threat of misery and destabilization, all this makes us, above all, bow our heads low before the efforts of millions of Poles, whose work, effort, patience and responsibility. (Tusk, 2011)

According to the prime minister, however, there was no reason for premature joy. Later in the exposé, he announced "blood, sweat and tears."

> Continuing the reform of the pension system is absolutely essential if we are serious about making Polish public finances truly secure. (…) This is a course of action that can guarantee us the security of public finances not for a year or five years, but lasting security. To put it in human language: this means financial security for precisely those who will depend on the pension system. If I did not have full conviction that the security of Polish pensioners today and in the future depends on this step, I would not propose it. (Tusk, 2011)

"Taking care of a healthy economy and sound public finances is an important part of building the security of Poland and Poles," E. Kopacz, who replaced D. Tusk in 2014, said in her exposé (Kopacz, 2014). Prime Minister Kopacz's exposé, like Prime Minister Tusk's, came at a time of economic crisis, when governments had to make unpopular decisions that hit Poles' household budgets. Hence the securitization measures in the face of the issue of the state of finances and the economic crisis. The state's and individuals' economic foundation is at risk, so there is a need for budget cuts, changes in the social security system, extending working hours, etc. If these actions of the government are not accepted, the state is threatened by a crisis on the scale of Greece.

It is difficult to assess whether the securitization succeeded, since a year later PO lost the elections, and the victorious Law and Justice party owed its success to rich social programs at odds with fiscal discipline. The 2015 campaign did not specifically address the problems of state finances. The Left spoke in general terms about "strengthening the state's economic security, including in financial terms" (Zjednoczona Lewica, 2015), the PSL, in a section entitled *Economic Security*, presented general slogans. PiS, except for energy, did not address the issue, and the incoming Prime Minister Beata Szydło (2015) placed economic security among her government's three security priorities, but pointed primarily to its energy dimension. Mateusz Morawiecki (2017), a former bank governor, also devoted little space to economic (financial) security, announcing generally: "Our program is the will to build a Republic of financially secure families" (Morawiecki, 2017). In the 2019 campaign, the topic was only addressed by KO in the program's chapter entitled "Economic Security." In it, she pointed to the record budget hole and recommended tax cuts, among other things: "The economic security of Polish women and men should be based on transparent laws and stable finances, not only on a macro-, but also on a micro-economic scale. The main pillars of microeconomic security for citizens will be the 'Lower taxes, higher pay' program" (KO, 2019).

"Poles have entrusted PiS with the task of building a Polish welfare state. A safe and modern state." – Mateusz Morawiecki said in his second exposé. – "The state's strategic investments will be a safety cushion against today's forecasts in the global economy" (Morawiecki, 2019). However, the securitization treatments seen under the PO government during the economic crisis are not here. The 2023 campaign referred to the dangers posed by high inflation and war:

> Inflation posed a threat to Polish families and consumers, as it caused fuel, energy and food prices to rise. The government, guided by the principle of social solidarity and state activity in the economy, decided to introduce special measures to protect Poles – the so-called anti-inflation, energy and solidarity shields,' boasted PiS. (2023)

The other parties did not address the subject.

2 Energy Security Discourse

As a separate topic in the category of economic security, energy security came to the fore only in the 21st century. Admittedly, PC wrote as early as 1991 about the state's duty to ensure energy security, and Prime Minister Waldemar Pawlak's 1993 exposé included an assertion that the government would control energy prices, but this did not frame energy in the context of security. But as early as 2001, PiS wrote in its program: "energy security is one of the most important factors in preserving Poland's independence." It went on to enumerate what needed to be changed. At the top of the requirements was diversification of raw material sources (PiS, 2001).

Thus, energy became one of the cornerstones of a free and secure state. In 2005, when plans for the Northern Gas Pipeline (Nord-Stream 1) were being concretized, the party advocated that Poland's foreign policy should ensure energy security as one of its overriding goals. All because of the threat coming from Russia.

> One of the primary goals of Polish policy must be to ensure the country's energy security. We can only achieve this by making Poland independent of Russian gas imports. (…) Our situation is aggravated by the fact that in matters of energy security we unfortunately cannot count on solidarity from the European Union. (PiS, 2005)

Kazimierz Marcinkiewicz (2005), who took over as prime minister, continued his party's narrative: "We will push for a joint European Union energy policy, but at the same time we will firmly defend Polish energy security. We will exercise all our rights in this regard." The statement was meaningful, as the prime minister indicated that Poland would block the German-Russian venture Nord-Stream 1. "Energy security guarantees will be a priority in my government policies" (Marcinkiewicz, 2005). His successor, Jarosław Kaczyński (2006), continued: "Today, external security revolves primarily around two problems, that is, economic security, particularly energy security, and security against terrorism." When assuming the office of president in 2005, Lech Kaczynski mentioned energy security in one sentence, in a general context, that "the threat to Poland's energy security and related events (…) are facts that cannot be disputed, cannot be denied" (Kaczyński 2005).

PiS has consistently raised the issue of energy security since 2001. Other parties at the time viewed energy primarily through the lens of energy prices, which the state should control or liberalize, or through the privatization of energy companies. The situation changed in 2007, after the crisis of the winter of 2005/2006, when Russia halted gas supplies to Ukraine. It was then that energy security began to be discussed by all groups.

"PO proposes to increase energy security by appealing to its own resources," the main rival of the PiS wrote in its program. PO did not refer to the Russian threat as the reason for the proposed changes. Its motivation was to modernize the sector for economic and environmental reasons: "We propose to enter the process of reorienting the energy industry to increasingly innovative technologies." Unlike the competition, party representants believed that this could be achieved through European solidarity: "We will be active in creating the principles of a joint EU energy policy so that our country's interests are guaranteed in it" (PO, 2007).

A similar faith in allied solutions was evident on the Left. The LiD program included a provision: "Ensuring energy security and the rational use of energy are among the basic directions of Polish economic policy." The way to implement the postulate was to be European solidarity, as Europe was struggling with the energy transition for environmental and security of supply reasons. The condition of the environment, or climate change, was the perspective through which the Polish left saw energy: "Poland should combat climate change and increase energy security among its priorities in the European Union. Focusing solely on the security of supply is a misguided approach" (LiD, 2007).

The difference in approach between PO and LiD and PiS shows that securitization can happen using different arguments and pointing to different solutions, even partially contradictory ones. The mechanism used in the case of PiS is as follows (Figure 17):

FIGURE 17 Securitization of the energy issues by PiS
SOURCE: OWN STUDY

For the left and center parties, the situation is somewhat different, although here, too, one can see a process of securitization of energy issues (Figure 18).

PiS pointed to a single culprit for the threat: Russia, while LiD and PO discussed problems with energy supplies and prices that are global in nature. They sought to resolve these issues through cooperation with the EU, which PiS distrusted, placing the emphasis on government action. The modernisation of energy-intensive enterprises and the introduction of new technologies were among the second set of measures proposed by LiD and PO, which PiS acknowledged but did not believe would have a significant impact.

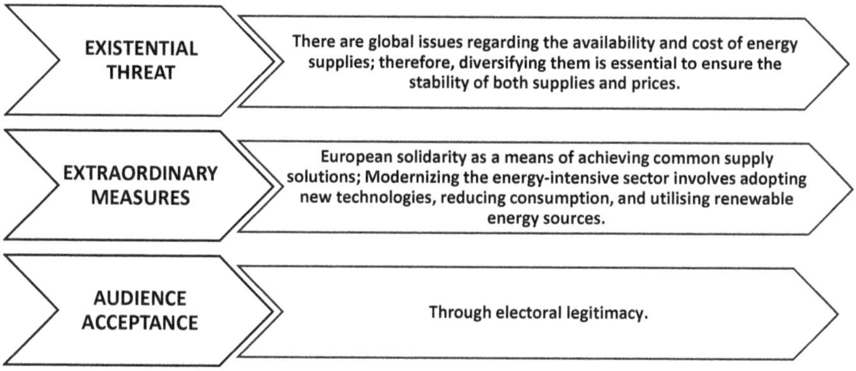

FIGURE 18 Securitization of the energy issues by center and left-wing parties
SOURCE: OWN STUDY

Unsurprisingly, in the party's subsequent program, energy was portrayed primarily through the lens of government initiatives. Meanwhile, Donald Tusk, upon assuming the Prime Minister's role, echoed the sentiments expressed in the election program:

> The most important element of economic security is energy security, which we understand primarily as a guarantee of uninterrupted supplies of energy carriers at acceptable prices with simultaneous concern for the environment. We will implement this policy within the national strategy framework, cooperating with partners from the European Union. (Tusk, 2007)

PiS's 2011 election program was rife with references to energy security. The issue was presented as part of foreign policy:

> Energy security is an essential aspect of security in general, treated on a par with military or trade issues. (…) We do not want to exacerbate relations with Russia against the background of conflicting interests in gas extraction, but we will vehemently oppose efforts to prevent Poland from exploiting its own resources. At stake here are our country's vital economic interests and security. (PiS, 2011)

The program also included references to domestic policies: special supervision and centralization in the energy field (establishment of the Ministry of Energy, oversight of energy companies). Shale gas was regarded as a vein of gold, enabling the diversification of energy sources and an expansion of the energy mix with autonomous raw material sources. Years later, after the project's collapse, it is difficult to ascertain how much political decision-makers and candidates for power believed in the idea's success. Nonetheless, it was discussed by all factions of the political scene.

> The project to launch shale gas exploitation is politically comparable to our country's membership in NATO. A Poland that becomes a gas exporter within a decade, that is able to stop the accumulation and pay off the existing debt, a Poland with surpluses in the budget to enable strategic investments and development is a secure Poland. (PiS, 2011)

PO presented energy through the lens of strategic transformation. It pointed out that "without a modern, efficient and secure energy sector, we will lose the opportunity to compete and strengthen our position in Europe" (PO, 2011). In her exposé, Ewa Kopacz said:

> "Our country will strive for the European Union to put into practice the idea of energy solidarity. We cannot accept a situation in which Poland and other countries in the region are condemned to the mercy and disfavor of external gas suppliers." She has also presented a hitherto unpopular approach in the PO regarding coal. "In Poland, coal is of strategic importance. One of the first decisions I have made is to speed up work on laws that will wisely combine three fundamental goals regarding Poland's energy security. First, we must protect Polish mining from unfair competition. Second, I will not relent in my efforts to make this industry finally profitable. Third, I cannot disregard the safety of every Polish family. Polish homes must be heated, and energy costs must not ruin household budgets." (Kopacz, 2014)

The belief in an energy-secure Poland was not shared by PiS, which stated in the very first pages of its 2015 election program that "energy must be available to all who need it (implicitly – it is not currently), while security of energy supply must be guaranteed by the State and the price of energy must not hamper the competitiveness of Polish industry." From this sentence alone, it is apparent that there has been a shift in the party's thinking from viewing energy through the prism of raw material sources and security of supply to understanding it as an increasing cost to citizens. The key point in the narrative has become coal, which PiS believes was the foundation of energy security in Poland. By fighting it, the European Commission was fighting the Polish state, destroying its security: "Renewable energy sources will find their place in the energy mix, but they will develop in Poland under market conditions and not ideological ones. Poland's energy security in the foreseeable future will be based on coal-fired energy, not renewable energy" (PiS, 2015).

These words should be read in a specific context. In 2015, PiS presented its election program in Silesia (a mining region of Poland), during a convention held in the center of Katowice, in the modern International Congress Center, which was built partly on the grounds of a former coal mine. Under such circumstances, the importance of mining to the region, which hosted party members and supporters, could not be underestimated. Defense of environmentally harmful solutions, in the Law and Justice program, took on the dimension of protecting state security.

A sharper opinion on solving energy problems seen from the perspective of the state of the environment was presented by Kukiz15: "The European Union's energy and climate policy is a purely ideological project for which there is no sensible scientific, economic or political justification. (…) European climate policy primarily hits the coal-based Polish energy sector" (Kukiz15). The party focused primarily on the economic impact of the energy policy changes proposed by the EU. It did not refer to access to raw materials, diversification of their sources or security of supply.

The PO in 2015 maintained its earlier stance on the inevitability of solving energy security problems through European cooperation. It also explicitly admitted that it underestimated the threat from Russia in this regard: "Tensions across our eastern border and attempts to use raw material supplies as a tool of political blackmail made ensuring energy security for Poles and the Polish economy our strategic goal" (PO, 2015). The Left devoted most of its attention to protecting the mining industry, claiming that "Poland's energy sector should be based on two pillars: renewable energy sources and coal-fired power generation. Renewable energy sources are an opportunity for millions of households that can produce energy for their own needs. Coal is a guarantee of energy security." The leftist coalition has prioritized preserving jobs in the mining industry, which it has argued is important for energy security (Zjednoczona Lewica, 2015).

When taking over the Prime Minister's portfolio, Beata Szydło (2015) raised the same themes included in PiS's election program. Her successor, Mateusz Morawiecki (2017), did not add anything new to his predecessor's words:

> Today coal is the foundation of our energy industry. We cannot and do not want to give it up. (…) But for our future generations, I would like to ensure

> that alternative sources of energy can also develop freely in Poland – not in the name of ideology, but where it is economically justified, where it will bring Poles benefits rather than costs. (Morawiecki, 2017)

This line was partially upheld in the Law and Justice program prepared for the 2019 elections. "The basis of the concept of energy security for Poland is the principle of energy sovereignty," the party reminded. "At the same time, we will remain steadfast in our position of opposing the construction of the Nord Stream 2 gas pipeline." The government's centralizing decision to establish a single oil company is also expected to be an important measure in this regard (ZP, 2019).

The energy security elements of the 2019 program differed from those in earlier programs. First, there was no mention of the existential threat that Russia has posed thus far. Second, there was a lack of other significant threats that could worsen the situation for Poland, which was successfully implementing diversification initiatives. Third, coal ceased to be the basis of national sovereignty and became one of many energy sources that could potentially be replaced by renewables in the future. Fourth, PiS adopted a more cost-effective approach to energy development, suggesting that it could not be prohibitively expensive. There were, of course, ongoing threads of central oversight of energy companies in the program, but it revealed PiS's new perspective on energy, which can be desecuritized due to its widely recognized importance.

KO saw energy primarily through the lens of the environment. The coalition understood energy security as relying on cheaper alternatives to the currently dominant and renewable energy sources. Koalicja Polska (PSL and Kukiz15) firmly stated that "Poland's energy security must not depend on the whims and blackmail of foreign governments" (KP, 2019).

> "Powerful investments in renewable and conventional energy or, finally, the construction of Baltic Pipe – a new gas supply corridor, which will also be a symbolic end to Poland's dependence on gas supplies from the East. This is certainly the normality expected by Poles in Poland's energy security," Prime Minister M. Morawiecki said in his second exposé (2019). – "For the first time in history, we are able to reconcile the three most important principles of our energy policy: energy security, price competitiveness and care for the environment."

The change in narrative presented showed that even conservative groups cannot address the problems of environmental pollution and the harmfulness of coal-burning energy. "Traditional power generation will still be important in our electricity system for a long time to come, but circumstances are changing. We once couldn't afford to develop renewables, and now we can't afford not to develop them" (Morawiecki, 2019).

The 2023 campaign, in the shadow of the war in Ukraine, compelled a return to energy issues from an economic rather than an environmental perspective. However, this return was not as detailed as it had been a decade earlier and did not exhibit the hallmarks of securitization. KO, Trzecia Droga, and Lewica did not address the issue. PiS dedicated a dozen pages of its program to it, recalling its diversification efforts and the strengthening of the state oil company. Yet, even this grouping admitted that the future lay in nuclear development and renewable energy sources, and that coal was only a "reserve." Konfederacja spoke of "the costs of the energy transition proposed by the European Union, which, according to this grouping, deprives Poland of its energy security, and thus, in part, of its political independence." In doing so, it pointed to the exact solutions as PiS – nuclear, coal and renewable energy. The incoming Prime Minister Donald Tusk (2023) unequivocally declared investment in nuclear energy.

Despite the renewed threat to supply stability, energy is an area that has also begun to undergo desecuritization processes in Poland. The issue of diversification of energy sources, even with the continued threat of instability of raw material supplies from Russia, has lost its importance with the solution of the problem, finding other, diversified, and stable sources of energy raw materials. The narrative of diversification aimed at increasing the share of green, clean, renewable energy in the energy mix has gained strength, but this discourse occurs primarily in the area of environmental (ecological) security and less frequently in economic security.

3 Social Security Discourse

Material security for citizens in a country undergoing economic transition, manifested by high inflation, unemployment, falling incomes, and a significant deterioration in living standards, has been an essential element of political programs since 1989. Attitudes toward the transition and proposals for reforming the economy were combined with demands for social protection for the most vulnerable. References were made to state actions, humanity, European standards, or human rights. This approach was present in all the groupings at the time. Moreover, as early as 1991, social security was explicitly mentioned as an important area of state activity (NSZZ "Solidarity") or even its system (Polskie Forum Chrześcijańsko-Demokratyczne).

At the beginning of the 1990s, social security was understood not only as ensuring the survival of citizens in material terms (work, pay, pension, access to free health care) but also as the foundation of a fragile social order and a young democracy. Its significance thus extended beyond the needs of individuals or even social groups; it became a matter of the functioning of the entire state, its order, system, and future. In 1991, Unia Demokratyczna pointed out that the observed "danger of increased frustration and discontent, which could consequently turn against the democratic transition and the market economy," was precisely related to the failure to provide basic social security. "We would have briefly enjoyed our regained political freedoms if we had failed to stop the economic catastrophe," Prime Minister Tadeusz Mazowiecki (1989) said in the same vein.

Jan Krzysztof Bielecki did not pronounce the word "social" once in his exposé, but he did draw attention to helping the most disadvantaged as a duty of the state (Bielecki, 1991). "Social expectations are linked to the state of the economy and social security. Regardless of the amount of funds we will have at our disposal, we consider the creation of a guarantee of the social minimum to be our primary task," Waldemar Pawlak declared in his exposé (1992). This was the first explicit reference to the concept of social security in an exposé after 1989. Józef Oleksy only mentioned "organizing and solving problems related to the sense of social security in the countryside" (Oleksy, 1995). Wlodzimierz Cimoszewicz spoke in general terms about respecting citizens' social guarantees. So, as can be seen, the prime ministers, unlike the parties from which they came, did not refer too readily to social issues in the security sphere, although they appreciated their importance for the stability of the state. The 1997 campaign was not groundbreaking in this regard – the concept appeared only sporadically. It did not appear with left-wing groupings. Instead, the PSL consistently used it, writing about: "The social security complex associated with a sense of social security" (PSL, 1997). Unia Wolności reminded that "the social Security System must give people a sense of elementary security" (UW, 1997).

Announcing planned reforms, Prime Minister Jerzy Buzek referred to security. "The pension system reform is the most important among the significant social sphere reforms. The pension system, towards which we will be moving, is to give pensioners guarantees of a secure old age," he said (Buzek, 1997). In the exposé, there was a lot about social safety nets for workers in reformed industries (such as mining) or agriculture. The very notion of social security, however, did not come up. In the 2001 campaign, social issues were seen from the perspective of security by two parties (PO and PiS): "Social assistance, education, health care and the provision of security are the basic duties that the state should provide to its citizens" (PO, 2001); "Millions of Poles today feel fear for their personal security, for their jobs, for jobs for their children" (PiS, 2001). The Left, which won these elections, did not take such an approach. Nor were the topics included in the exposé of new Prime Minister Leszek Miller (2001). Marek Belka did refer to social issues, but without a security context. The 2005 campaign did not differ significantly in this regard from others after 1995. Two parties addressed the topics: We aspire *to* "rebuild the Polish State in such a way that it begins to help and serve Poles by providing security by guaranteeing the payment of pensions for life" (Ruch Patriotyczny, 2005); "We advocate the implementation of the constitutional principle of a social market economy, which has proven itself in the development of societies in European Union countries, providing their citizens with prosperity and social security" (SLD, 2005).

Kazimierz Marcinkiewicz's expose included the issue of minimum wage, but it was not securitized. Instead, in the 2007 campaign, the issue appeared with three parties, two of which (PO and Lewica i Demokraci (LiD) coalition) referred to the concept of flexicurity (a combination of the words *flexibility and security*): "It involves combining labor market flexibility measures and the use of new security tools. It is a transparent system of social security and guarantees of basic labor rights" (PO, 2007). The PSL returned to previously advocated social security proposals. However, the Left had the most to say in this area. It began with a critical diagnosis of the situation: "Despite the favorable economic situation in Poland, there has been no significant improvement in social security to date. Meanwhile: Conditions now exist to provide Poles with a sense of long-term social security after years of sacrifice. (…) We want to free people from fear," it declared (LiD, 2007).

Prime Minister Donald Tusk has dreaded the specter of social spending cuts if the budget struggles to stabilize: "Only in this way can we permanently remove the threat of renewed tax increases and sudden cuts in social spending or public sector wages," he argued (Tusk, 2007). PiS, which in 2007 did not refer to the issue of social security, in 2011 presented it as an issue of civilization, quality of life, part of economic security. SLD presented social security as one of the duties of the state, referred to the reform of the pension system and job security: "The state, washing its hands of its most important duties: responsibility for social security, for everything that makes us feel safe, causes many Poles to see their lives in terms of a ruthless struggle for existence" (SLD, 2011). The PSL spoke of providing citizens with abundance: "The people's priority is first and foremost to improve the material living conditions of Poles" (PSL, 2011).

PO declared laconically: "We will guarantee pension security" (PO, 2011). Donald Tusk's second exposé, based entirely on economic issues and the economic crisis, began with the grim statement that: "millions of people accustomed to peace, prosperity and security today face the threat of misery and destabilization" (Tusk, 2011). In further parts of the exposé, the prime minister declared that "all the government's efforts must be directed toward the security and well-being of each individual, the nation, the state." Every subsequent government action described had either economic security (of the state and citizens) or social security (of citizens and social groups). This is how the Prime Minister explained the plan to extend the retirement age:

> We need to think not only in terms of the upcoming budget and this term, but also such financial stability that will enable the stable payment of pensions to todays and future pensioners. To achieve this, it is necessary to raise the retirement age. (...) This means financial security for precisely those who will depend on the pension system. (Tusk, 2011)

What we have here is a securitization process that was started, which was not realized due to the change of government in 2015 and the withdrawal of the proposed reform. Public acceptance was lacking. Nevertheless, it was security that was the appeal to help citizens accept a difficult decision (see Figure 19).

In the 2015 campaign, the theme of security of future pensions returned. Nowoczesna diagnosed and postulated: "The modern pension system should be supplemented with a capital part that will increase pension security" (Nowoczesna, 2015). PO – on the issue of pensions – insisted on the position presented earlier by announcing: "We will defend the pension reform, gradually extending the length of working life, in the name of dignified old age security" (PO, 2015). Zjednoczona Lewica proposed: "It is necessary to guarantee rental security not only for landlords, but also for tenants" (ZL, 2015). Another

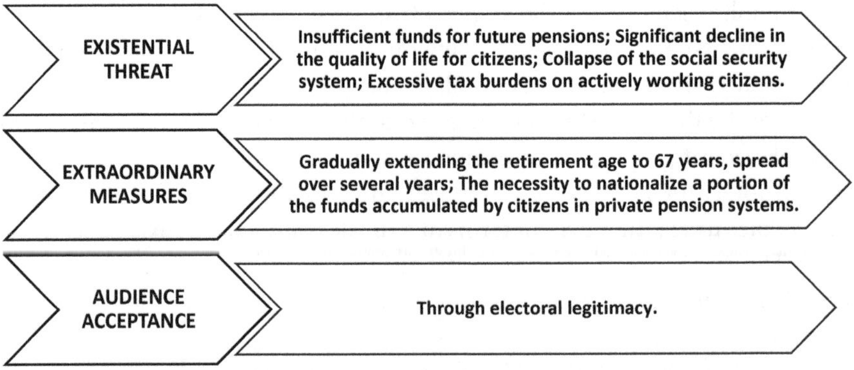

FIGURE 19 Securitization of the reform of pension system in the political discourse of D. Tusk
SOURCE: OWN STUDY

leftist party, Partia Razem, focused on work and forms of work: "Sickness, pregnancy or the loss of a job for many Polish citizens and nationals means the loss of an elementary sense of security. Poland is an increasingly wealthy country, but stable and secure work remains the privilege of a few" (Partia Razem, 2015).

However, PiS had the most to say on the topic of social security in 2015. The first part of the diagnoses and demands concerned the pension system. "A comparison of the current level of pension security in Poland with solutions in other EU countries provides an essential rationale for emigration." The next part of the program included references to employment stability and the condition of the family: "fertility growth is not fostered by high unemployment, **lack of job security**, limited access to housing and care facilities" (PiS, 2015). Representing this party, Prime Minister Beata Szydło (2015), announcing in her exposé extensive social programs and lowering the retirement age, did not refer to security to defend these government decisions. The key words here were: justice and family. Mateusz Morawiecki spoke in the same vein: "The government will pursue a policy of supporting Polish families. Development and social issues are the same" (Morawiecki, 2017). In the 2019 campaign, PiS (as coalition Zjednoczona Prawica) returned to the issue of retirement security, postulating: "Saving for retirement will become safer and more transparent. This is to be served by the introduction of Employee Capital Plans" (ZP, PiS, 2019).

KO declared: The 13th pension will become a guaranteed benefit – to make seniors feel more secure (KO, 2019). Koalicja Polska announced broad implementation of social programs. The Left focused on issues of employment conditions: "We will ensure constitutional labor rights for all employees, including the right to safe and healthy working conditions, to social security, to a minimum wage" (Lewica, 2019).

In his second exposé, Prime Minister Mateusz Morawiecki referred to the notion of a welfare state, which the government stated as its goal for construction in Poland, equitable development (he announced the maintenance of social transfers), and support for families. However, there was no context of security here. Social transfers became a permanent fixture in the government's program and gained social acceptance, so there was no need to justify them in an extraordinary way. Nor did the 2023 campaign securitize social issues. The parties stuck to the simple conundrum that social security is important. So did both prime ministers – M. Morawiecki and D. Tusk.

Social security has been a consistent feature of election programs and political speeches since 1989. For most of this time, however, it has primarily been viewed through a political lens, largely due to the general consensus on the state's functioning and responsibilities in this area. We are only addressing securitization in the context of the pension reform announced by Donald Tusk's government. However, it must be regarded as having failed because of the lack of public acceptance of the proposed solution, as expressed through voting in parliamentary elections.

4 Health Security Discourse

Health security, as a concept, did not appear in Polish political discourse until the end of the 20th century, nevertheless that health topics were considered an important issue in the 1990s. The term itself first appeared in Unia Wolności program in 1997, as the title of one of its chapters, "Health Security: Reforming the System to Reduce Infant Mortality, Premature Mortality, Cardiovascular Disease" (UW, 1997). The UW's predecessor, UD, in announcing health sector reforms in 1993, recommended: "reforms in the health sector are indispensable, and their implementation should begin immediately, with a margin of safety, due to the special matter" (UD, 1993).

This term did not appear in the first exposés of prime ministers, although they talked a lot about health. Tadeusz Mazowiecki diagnosed: "We are aware of the dangers threatening the biological existence of the nation" (Mazowiecki, 1989). "We all know about the disastrous situation of the health service," Jan Olszewski (1991) echoed in an equally alarmist tone. Waldemar Pawlak promised: "We are not going to back down from our obligation to create organizational and financial conditions for health care and social welfare" (Pawlak, 1992). It was not until Hanna Suchocka became the first prime minister to use the term health security: "Public health care must provide each of us with a minimum of health security and guarantees of receiving help in an emergency" (Suchocka, 1992).

"Only a healthy society can build a strong and prosperous Poland. Unfortunately, the state of health of Poles fills one with concern" – diagnosed W. Pawlak (1993), while heading the new SLD-PSL government. Józef Oleksy (1995) focused on the technical aspects of health care reform, Włodzimierz Cimoszewicz (1996) spoke of the reform as a civilization challenge. Jerzy Buzek (1997), taking office as prime minister of a government planning four major reforms, including health care, declared: "There will be a thorough analysis of the needs for equipping hospitals and emergency care with equipment to increase patient safety" (Buzek, 1997). Leszek Miller (2001) and Marek Belka (2004) did not present health issues from a security perspective. In the 2005 campaign,

PO was the only party to return to the notion of health security: "the health security of citizens requires giving a rigorous framework to the entire health sector" (PO, 2005).

"My government considers it crucial to ensure health security for every citizen," new Prime Minister Kazimierz Marcinkiewicz (2005) declared after PiS won. In the 2007 campaign, health issues from the perspective of security were presented by two parties, including the PSL, which was consistent in this regard: "We will guarantee all necessary benefits for maintaining and improving health within the framework of existing insurance" (PSL, 2007). "We plan to repair the health care system, but we are not offering Poles a revolution. Relentless reforms and counter-reforms have led in no small measure to a loss of a sense of health security," Prime Minister Donald Tusk asserted in his exposé (2007). PiS spoke in similar terms about reforms in the sector in subsequent elections: "The program is aimed at creating a friendly and trustworthy health care system, which will ensure equal and equitable access to publicly funded health care services and increase the level of health security of patients and improve the health status of Poles" (PiS, 2011).

The SLD has used health security as an argument against reform:

> Shifting responsibility for the public's health to local governments or directly to for-profit business entities is the reason for the dramatic reduction in the sense of health security. (…) The government's obsessive drive to commercialize and privatize all health care facilities has led to a dramatic reduction in the health security of Poles. (SLD, 2011)

"Health care is one of the most sensitive areas of social life. The security of citizens largely depends on it," PO declared (2011).

In the 2015 campaign, four parties referred to the health theme. Nowoczesna pointed to Poles' fear of illness and the need to get to a doctor, PO focused on the safety of the services offered rather than the threat of a lack of availability, the Lewica spoke of the dangers to women of not having access to abortion, and PiS spoke of health insecurity as a reason for emigration. "Today, the attention of most Poles is focused on the fear of not getting sick, and the health care system does not provide security. We will strive to change it to provide every Pole with an adequate, equal level of security" (Nowoczesna, 2015);

> Strengthening empathy is the basis for security and trust in the relationship between the patient and the treatment team. (…) We want to abolish

> the so-called anti-abortion law. Pregnancy termination procedures are performed in Poland underground, often in conditions that are dangerous to the woman's health. (ZL, 2015)

"Every Polish citizen is supposed to feel safe when he needs medical help. He is supposed to know that this help will be given to him quickly, that he will be safe," Prime Minister Beata Szydło (2015) said in her exposé after PiS won the elections, justifying the suspension of hospital privatization.

Standing before the Parliament for the second time, Mateusz Morawiecki (2019) again did not securitize the issue. He could not have known at the time that the real breakthrough in the perception of health as a security issue would come with the COVID-19 pandemic in March 2020. The securitization of the pandemic issue was a phenomenon strongly present in Polish and global political discourse, and would certainly be visible in the parties' electoral programs if parliamentary elections were to be held in 2020–2021.

Every session of the Parliament from March 2020 to the end of the year dealt with the COVID-19 pandemic, including a special debate titled *Information of the Prime Minister on the state of the country's preparedness for the rising wave of SARSCoV-2 coronavirus infection and the COVID-19 disease caused by it*, held on October 21, 2020.

In Prime Minister Mateusz Morawiecki's speech initiating the debate, there were a lot of references to security issues and military comparisons. "There are no places in a widespread epidemic situation that are completely safe." Therefore, it required extraordinary measures:

> We implemented new strictures, the obligation for everyone to wear masks. (…) We must forcefully adapt such methods that will lead to the fact that we will study confidently remotely and work, but we will protect our seniors, our most vulnerable. (…) The restrictions must be strictly enforced. (Morawiecki, 2020)

And he announced: "these measures will evolve, because the epidemic is dangerous by being unknown, an uncontrollable, unpredictable phenomenon."

The government's efforts to fight the pandemic were also extraordinary: "we bought, ordered, delivered, distributed, increased (the number of beds, respirators, masks, fluids, gloves, thermometers, tests, disinfection stations)," the prime minister enumerated. His Cabinet's actions were also about saving the economy. "We reacted with lightning speed, we reacted with an anti-crisis shield, a liquidity cushion for entrepreneurs," boasted the head of government.

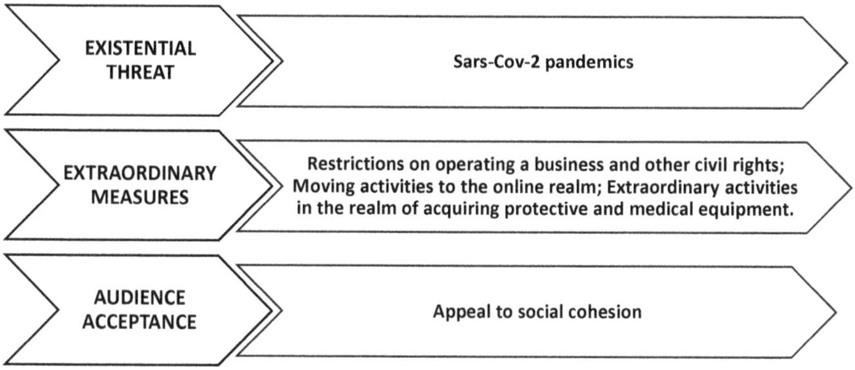

FIGURE 20 Securitization of the Sars-CoV-2 pandemic
SOURCE: OWN STUDY

> But I warn, he announced, Ignoring the epidemic in the name of the value of defending the economy could lead to something worse. Let's reject it and take the middle road, it is the safest one. (…) We must set ourselves up for a longer fight. Together we will come out of this fight victorious. (Morawiecki, 2020)

The speech clearly demonstrates the securitization efforts that have accompanied the government's actions since the beginning of the pandemic. An unusual, unprecedented, and unpredictable threat requires extraordinary measures (strictures, online schools, support for the economy, equipment purchases outside the public procurement system). The public had no choice but to accept it; there were severe penalties for violating the restrictions imposed unlawfully on many occasions. However, there was also no shortage of accountability for the health and lives of those exposed to more severe diseases. The securitization process is presented in Figure 20.

Members of Parliament participating in the debate were divided into a group supporting the government's actions (representatives of Zjednoczona Prawica) and the opposition criticizing the prime minister. Table 9 presents the key threads of the debate from the perspective of securitization of discourse.

As seen from Table 9, the language of military analogies reigned supreme in the parliamentary debate on the fight against the pandemic: war, victory, winning, losing, defeat, threats, enemy, weapons, fighting. There were also references to emergencies such as fire. Safeguard, counter, combat, ensure, protect, control, oblige, extinguish (fire) were the most popular verbs referring to action in the area of combating the coronavirus. There were analogies made to

TABLE 9 Debate on the Sars-CoV-2 pandemics in terms of security

Zjednoczona Prawica	Opposition
"The increments of new beds are all ahead of the number of new illnesses. This is our security guarantee. We are creating a safeguard. We have prepared a package of legislative solutions, which has becomę another weapon in the fight against the pandemic, which will allow us to give greater safety guarantees to staff, greater financial incentives, while at the same time taking care of patient safety." (Adam Niedzielski, Minister of Health)	"Instead of fighting a pandemic, you are fighting the truth. (…) Today the state is barely panting hooked up to a respirator." (Borys Budka, KO) "Are you ready for the worst? Have you done everything you are obligated to do so that we have a food and water supply, road safety, protection from robbery, a supply of energy resources, a supply of fuels such as gasoline, diesel fuel, a supply of system heat and coal individually; so that there is access to our money; so that transportation services work; so that utilities function; so that there is protection against speculation and, in emergencies, a system of product rationing; so that emergency services function; so that there is continuity of all state services?" (Włodzimierz Czarzasty, Lewica)
"Please do not scare citizens with a COVID war." (Teresa Wargocka, PiS)	"The government is incapable of providing security for its citizens and is unfit to govern." (Krzysztof Gawkowski, Lewica)
"You don't have to be afraid, just take precautions." (Barbara Bartuś, PiS)	"The government has lost control of the pandemic, the exercise of functions designed to take care of security." (Władysław Kosiniak-Kamysz, PSL)
	"You have wasted 6 months and left Poland defenseless." (Marek Sowa, PO)
	"Today we are dealing with a serious fire. You played with fire and now our house is on fire, the whole system is no longer efficient. As the opposition, we will give you water, but you, Mr. Prime Minister, must be the firefighter today who puts out the fire, not starts it." (Mateusz Bochenek, PO)

SOURCE: OWN STUDY BASED ON THE DEBATE MINUTES

the extraordinarity of the situation and the need for non-standard solutions and quick action even against existing procedures, indicating the securitization of the issue. Health issues seemed to be an important part of the 2023 campaign, even if the pandemic had been defeated. Meanwhile, references to health issues in the context of security were sporadic. The pandemic appeared only generally in the program, which PiS wrote laconically that "epidemic and drug safety were priority aspects of health security" (PiS 2023), while Lewica addressed "patient safety and treatment safety." Mateusz Morawiecki mentioned increasing spending in general, and Donald Tusk talked about women's health in the context of abortions. The issues were not securitized.

5 Migration-Security Nexus Discourse

Migration issues in Polish political discourse appeared different in the 20th and 21st centuries. The first half of the analyzed period focused on the problems of the Polish diaspora scattered around the world, the repatriation of Poles, especially from across the eastern border, and the resolution of their grievances, as well as the situation of national minorities in Poland. Thus, the issue of migrants was not dominated by the theme of "others, strangers, threatening, hostile," but rather by a focus on "their own, close, dear, welcomed with open hands." All parties spoke with one voice. The prevailing rhetoric emphasized duty, solidarity, repayment of debt, and reparations. A summary is presented in Table 10.

Political representations of minorities living in Poland, such as Mniejszość Niemiecka (MN, German Minority), also spoke out: "Securing opportunities for full and equal participation of members of the German minority in the political, social, cultural and economic life of Poland. Creating conditions for securing the cultural, educational, religious and economic needs of the German minority" (MN, 1991) or Białoruski Komitet Wyborczy (Belarusian Voting Committee) speaking about the rights of Belarusians in the Bialystok region.

As if responding to these demands, Tadeusz Mazowiecki said: "Poland is a country – a homeland – not only of Poles. We live on this land with representatives of other nations. We want them to feel at home here, to cultivate their language and enrich our community with their culture" (Mazowiecki 1989). Jan Krzysztof Bielecki (1991) did not mention the issues of minorities or Polonia at all, Jan Olszewski (1991) was the first to speak about the emigration of Poles: "We want a Poland from which young people will not have to emigrate, because here they will find in their homeland the opportunity for a life start"

TABLE 10 Migration policy in the political program in 1990–1991

"The primary objective of Polish foreign policy is (...) to defend Polish citizens residing outside its territory. Nearly one-third of Poles live outside Poland's borders." (PSL, 1990)
"Guaranteeing that the rights of minorities living in Poland are respected, we demand, on the basis of reciprocity, similar treatment for Poles living together with Lithuanians, Belarusians and Ukrainians outside Poland's borders." (PSL Solidarność, 1990)
"Poland accepts the right of Polish citizens of German origin to maintain their cultural identity on condition that they maintain their loyalty to the Polish state. We demand on analogous terms the treatment of Poles residing in both German states." (PSL Solidarność, 1990)
"We consider it our duty to support with all our strength the Poles living in neighboring countries in their religious, national and cultural aspirations." (ZChN, 1990)
"Poland's foreign policy should demand that Polish ethnic groups living in the territories of other countries enjoy the same rights as the various minority groups living in Poland." (Polska Partia Socjalistyczna, 1990)
"Defense of citizens residing outside the territory of Poland." (PSL, 1991)
"An issue that requires careful attention is the status and fate of Poles living in the USSR. We consider it a patriotic duty to co-create the conditions for a dignified existence for Polish citizens living there, as well as to fully open Poland to those wishing to return to their homeland permanently." (Stronnictwo Demokratyczne, 1991)
"Respect the rights of national, religious and cultural minorities, ensure that they are able to participate fully in social life, and defend groups and individuals from discrimination on the basis of world views." (Unia Demokratyczna, 1991)
"We will make it possible for all those who feel Polish to return to their country, especially the cruelly experienced compatriots from the Soviet Union." (Partia Wolności, 1991)

SOURCE: OWN STUDY

(Olszewski, 1991). He devoted an entire paragraph to the rights of Poles outside the country: "The Polish state has a moral obligation to protect the rights of Polish minorities and to secure the interests of these minorities in agreements under international standards" (Olszewski, 1991).

"I would like to assure our compatriots scattered around the world that their fate is very close to us, and concern for the welfare of Poland has a special value for us." – Waldemar Pawlak (1992) said in his exposé. "In relations with

Russia, Ukraine, Belarus and Lithuania, we are particularly interested in issues: (...) the situation of Poles living beyond our eastern border" – declared Hanna Suchocka (1992). The theme of concern for compatriots abroad dominated the 1993 election campaign. For the first time, however, there were postulates of restricting entry rights to Poland or changes in policy toward national minorities. Samoobrona proposed: "National minorities must enjoy full rights, but not privileges" (Samoobrona, 1993). Jarosław Kaczyński warned: "Border control of a certain category of foreigners must be improved and strengthened, making it possible to limit the influx of criminals from abroad" (PC, 1993). "Strict control of border traffic is necessary both for internal security and to protect the Polish market" (Katolicki Komitet Wyborczy Ojczyzna, 1993). In the same style in 1997, ROP proposed: "We will seal the borders and stem the flow of criminals from the East" (ROP, 1997). Before that, however, Waldemar Pawlak said: "We will also not forget those Poles who live in countries poorer than Poland" (Pawlak, 1993). "Our priorities include effective support for Poles living abroad in securing their rights and contacts with their homeland," declared Jozef Oleksy (1995). "Approximately one million citizens of our country belong to national and ethnic minorities. A free and democratic state guarantees them full freedom to cultivate their traditions, cultural identity, language, and non-discriminatory treatment," noted Wlodzimierz Cimoszewicz (1996). As can be seen, over the first decade after 1989, the approach to the issues was quite consistent and did not arouse political differences. Standing at the head of the AWS-UW government, Jerzy Buzek spoke in a similar tone to his predecessors: "I declare an open-door policy for our compatriots from the West and the East" (Buzek, 1997).

The processes of integration into the European Union began to leave their mark on the approach to migration issues. On the one hand, Poland's eastern border was to become the external border of the European Union, which brought new responsibilities; on the other hand, the approach to refugees or national minorities had to change in a united, multicultural Europe. The country was responsible for fulfilling its obligations under the common security policy and the principles enshrined in the Schengen agreements. "Sealing the borders, however, could not mean isolation from its eastern neighbors," SLD wrote in its program in 2001.

The issue of migration and national minorities, including Polonia abroad, was raised in the 2005 elections only PiS and Polska Partia Narodowa (PPN). PiS supported Polonia and expressed concern about the excessive emigration of Poles outside the homeland. PPN, on the other hand, saw migration as a threat to Poland's identity: "We are against the influx of emigrants from all over the world, who treat Poland as a transitory country, a haven for financial abuse, contribute to the rise of banditry, take jobs away from Poles, and do not assimilate" (PPN, 2005).

The issue of migration and Polonia was addressed by both prime ministers of the two-year term of the Sejm. Kazimierz Marcinkiewicz complained about Poland's demographic problems, however, he did not see the solution to this issue in potential immigrants:

> The possible wider opening of Poland to immigrants will only partially solve the problem of the threat to economic development and the inefficiency of the pension system. In the process, it will create the dramatic problems of mass immigration that we can see today in Western European countries. (Marcinkiewicz 2005)

Jarosław Kaczyński declared:

> We must make a breakthrough in our relations with the Polish community. (…) This is a great resource. Millions of Poles, including those with good social standing, live outside our borders. Among them are also those who would be ready to return here, bringing many qualifications that are rare in Poland. (Kaczyński, 2006)

In the 2007 campaign, PiS wrote: "Responding to the needs of the latest labor emigration, we launched the program 'Closer to Work, Closer to Poland,' which aims to provide care for Poles leaving Poland to work in other EU countries" (PiS, 2007). The issue also appeared in the programs of the PO and LiD: "The Republic must take special care of the Polish labor emigration of recent years" (PO, 2007); "Encouraging conditions should be created for the return of Poles who left 'for bread'" (LiD, 2007).

The 2007 programs also raised the issue of policy toward possible immigrants. The left postulated: "A strong Poland in a strong Europe and a globalizing world also means participating in the creation of a common migration policy based on unified rules for the admission and employment of third-country nationals" (LiD, 2007). Completely different positions were taken by the UPR: "We consider it unacceptable for national minorities to have powers that place them above the rest of the Polish population, such as electoral preferences granted unjustifiably to the German minority" (UPR, 2007).

When taking office as Prime Minister, Donald Tusk referred only to the issue of the conditions for Poles to return to their homeland: "It is necessary to create incentives for temporary migrants to return to the country" (Tusk, 2007). But already in the following campaign, the issues of illegal migration and

Poland's migration security featured extremely prominently. The SLD devoted particular space to them:

> "It is necessary (…) to effectively eliminate illegal immigration. Continuous improvement is required of the integrated management system for the security of the Schengen external border. Population migration is an inevitable phenomenon – this process will increase. It is important that it is monitored, and that the state's migration policy meets our broad interests." Immediately, however, for the sake of balance, it was declared that: "Foreigners who come to us legally, respecting our laws, should fully enjoy the legal protection to which they are entitled, and their customs and religion should be respected, as long as they do not violate our legal order." (SLD, 2011)

In his second exposé, Donald Tusk (2011) did not refer to migration issues. But already his successor, Ewa Kopacz (2014), had to confront the issue in a special debate on the European Union's migration policy and its impact on the Polish government's stance in the context of the potential admission of refugees to Poland (*Information of the Prime Minister on the migration crisis in Europe and its repercussions for Poland*). Amid the humanitarian crisis Europe was experiencing as it was "flooded" with waves of migrants from the Arabian Peninsula and Africa. The issue of the forced relocation of migrants from camps located in southern European countries to other EU countries also ignited emotions in Poland, which was supposed to take in a group of 2,000 people, according to European Commission calculations.

All participants in the debate securitized the problem, although in the case of opponents of accepting refugees it was a direct action (migrant/refugee = threat/terrorist), and in the case of supporters of the humanitarian approach – a reaction to the attitude of the first group (assuring that while security is an important criterion here, there is nothing to worry about because it is intact).

There were three positions in the debate. The first, which was completely opposed to accepting any migrants, including refugees, was presented by PiS and the KORWIN Party. They explained this position entirely in terms of security considerations. The second, intermediate stance, presented by the government parties (PO and PSL), did not deny that accepting migrants was a security issue but assured that it was not a threat. Both attitudes thus securitized the migration issue, admitting that extraordinary measures should be taken to protect the security of Poles. Opponents of accepting migrants proposed openly breaking EU solidarity and refusing to accept refugees altogether. Those in favor of the moderate option advocated applying extraordinary security procedures (identity verification, special centers, surveillance, etc.) while accepting a limited number of migrants. Only the Left primarily presented a humanitarian approach, pointing out that there are as many good and bad people among migrants as in other social groups, and that Poland, whose citizens have migrated around the world, has experienced hospitality in many countries and has a debt to repay. Left-wing groups refused to view the migration issue solely in terms of security but, nonetheless, had to refer to such an approach due to the perspectives of other participants in the debate.

The government side (moderate position) was represented in the debate by five people: the prime minister, the foreign minister, the minister and deputy minister of the interior, and the deputy foreign minister for European affairs. Explaining the government's position, Prime Minister Ewa Kopacz referred to solidarity and the humanitarian dimension of the plan to welcome refugees to Poland, and recalled that the migrants previously accepted were not terrorists:

> In the 1990s, when Poland was a much poorer country than it is now, we took in 86,000 Chechens. Have you observed terrorists on every street corner? Have you noted any situations that would threaten the security of Poles? So I ask all those in the Sejm: Can we afford today to make a gesture of solidarity to those who are actually fleeing their country because they are afraid of losing their health or their lives? (Kopacz, 2015)

She further enumerated the security measures taken: "Today our borders are secure. We are ready to implement a well-thought-out, comprehensive plan (…) first, it is about separating economic migrants from refugees, and second, sealing the borders" (Kopacz, 2015). Speaking after the Prime Minister, Foreign Affairs Minister Grzegorz Schetyna began by accusing the issue of being manipulated for political purposes: "We are being held hostage by irresponsible politicians who scare us with a flood of refugees and the beginning of the end of European civilization, a new march of barbarians to lead our continent to its downfall" (Schetyna, 2015). He argued that it is Poland's duty to take part in solving the migrant crisis:

> It's about sealing the European Union's external borders. It's about lasting stability in the regions from which refugees are arriving in Europe. Turning our sides or backs on our partners, on the countries that are targeted for emigration today, goes against the spirit of solidarity and may cost us the loss of solidarity on those issues that we particularly care about. (Schetyna, 2015)

Interior Minister Teresa Piotrowska focused on assurances about the country's readiness to protect the safety of Poles: "Poland has strengthened procedures for the interaction of services and institutions in case of a possible influx of large numbers of refugees." According to her, however, there was no cause for concern: "We are not seeing a sharp increase in the number of illegal crossings of our border due to the migration situation in southern Europe. Individual attempts at group smuggling of migrants into Poland are revealed. Border traffic remains at a level similar to the same period of 2014" (Piotrowska, 2015).

According to the head of the Interior Ministry, Poland was prepared for the relocation:

> In all the measures we take, security is a top priority. That is why it is so important to be able to check and to be able to refuse any person who could pose any threat. The adopted scheme of operations prior to the possible transport of persons to Poland provides for their unequivocal identification and fingerprinting. The provisions of the Law on Granting Protection to Foreigners on the Territory of the Republic of Poland provide for appropriate mechanisms to prevent threats. Thus, we have developed appropriate procedures and mechanisms. (Piotrowska, 2015)

Deputy Interior Minister Piotr Stachańczyk further explained: "The resettlement procedure is that we go to UNHCR centers and select them there. We can talk to these people, we can check them through our services, through local services. These 900 people will be an even more vetted group" (Stachańczyk, 2015).

Rafał Trzaskowski, deputy head of the Foreign Ministry in charge of European Affairs, was taking a conciliatory attitude when explaining the government: "Above all, a rational approach is needed at this point. It is an approach that is between two extremes – on the one hand, the voices that say that there is really nothing to negotiate and absolutely everyone must be accepted, and on the other hand, the claim that actually this is not our problem" (Trzaskowski, 2015). He reiterated the words of previous speakers, stressing how many safeguards have been taken to ensure that the security of Poles is not threatened: "A holistic approach is needed: securing the borders, distinguishing between refugees and economic migrants and vetting them for safety. It is also important to ensure that the Polish government has control over migration flows" (Trzaskowski, 2015).

Rafał Grupiński, representing the PO club, spoke in a similar tone. He spoke of "a great humanitarian disaster and intra-European solidarity," and he decried the identification of migrants with terrorists: "Please remember that political refugees, if only from areas of Syria, are refugees who are fleeing the Islamic

State. Building, creating such an atmosphere around this, that Polish newborns will be blown up by terrorists, is completely irresponsible" (Grupiński, 2015). After him – in the rest of the debate, several more proponents of the moderate approach spoke. Stanisław Żelichowski of the PSL said: "the first step we should consider is to do everything so that these people have somewhere to return to. The idea is, on the one hand, to provide security and the opportunity to find good jobs there, and on the other hand, to create such a Marshall Plan for these parts of the world, so that they can develop." His party colleague John Godson admitted that "our worries and fears about the unknown are normal today. I also have such fears sometimes. However, I want to say that fear is a bad advisor."

The pool of opponents of accepting refugees in Poland was opened by PiS leader Jarosław Kaczyński. Not only did he draw attention to the threat posed by potential refugees about to arrive in Poland, but above all he stressed that this was happening because of a decision by Germany and other European countries, not a sovereign decision by Poland:

> Does the government have the right, under external pressure and without the consent of the people, to make decisions that could have a negative impact on our lives, on our everyday life, on our real sphere of freedom, and finally, on our security. (...) This is not about accepting this or that number of foreigners. There is a serious danger that a process will be set in motion that will look more or less like this: first the number of foreigners increases rapidly, then they do not respect our laws, our customs, and then they impose their sensitivities and their requirements in the public space in an aggressive and violent manner. If someone says this is not true, look around Europe. (Kaczyński, 2015)

Admittedly, PiS president stated in the final part of the speech: "we can help refugees, but in a way that is safe for Poles," which meant more or less that humanitarian aid can be sent to Syria.

Other representatives of PiS, Solidarna Polska, Polska Razem and the KORWIN party spoke in a similar tone. "You demand that Poles renounce common sense. Your naiveté resembles the naiveté already known from history, when in 1938–1939 it was said that Hitler is a good master who can be controlled. The Islamic State is such modern Nazism," said Patryk Jaki (Solidarna Polska).

> Poland was strong because it was tolerant, but also because it had rulers like Jan III Sobieski, who also defended Poland far beyond its borders, to

> keep threats from the expansion of other cultures far from our borders. (…) 75% of these alleged refugees are strong men who come to do jihad here in Europe. (Zbigniew Girzyński, independent, formerly in PiS)

"When the leaders of the Islamic State say they will cut off heads, they later cut them off. Similarly, when they say they will send fighters to Europe, who are not fleeing war at all, but want to bring war to Europe, they want to bring this war to Poland," argued Przemysław Wipler (Partia KORWiN).

The debate also included arguments relating to economic and social security:

> It is well known that many Poles today live in poverty, in squalor, many people do not have jobs, and they are also most hurt by economic issues in the context of what awaits these future refugees. Is it true that the Polish government will offer free housing to all immigrants? (Andrzej Kania, Polska Razem)

Speaking on behalf of SLD Tadeusz Iwiński took a humanitarian approach to the issue of migration:

> "People fleeing war and political persecution should be helped. Poles, victims of persecution and war, have been taken in by third countries many times in history, so Poland should take in people fleeing war, while taking care of their social integration and protecting public safety," Iwiński argued.

"I want to believe that Poland is such a stable and prosperous country that even without EU assistance it would be able to take in the unfortunates fleeing death and torture," said Wanda Nowicka (SLD). "Not all Syrians and refugees are terrorists. How much one must have no respect for human dignity to see a potential terrorist in every person who is in a refugee camp today – both a child and a woman" (Ryszard Kalisz, SLD). "We are discussing today the admission of several thousand desperate, intimidated people fleeing war and persecution. What terrible disbelief we have in the strength of our own society if we think we are threatened by 10,000 refugees," argued Piotr Bauć (Ruch Palikota).

As can be seen, the issue of migration, particularly the reception of potential refugees in Poland, was a challenging topic for politicians in the middle of the second decade of the 21st century. Right-wing groups held an unequivocally clear position, securitizing the issue and warning of a flood of terrorists. The government (PO and PSL) and some MPs from left-wing parties adopted an

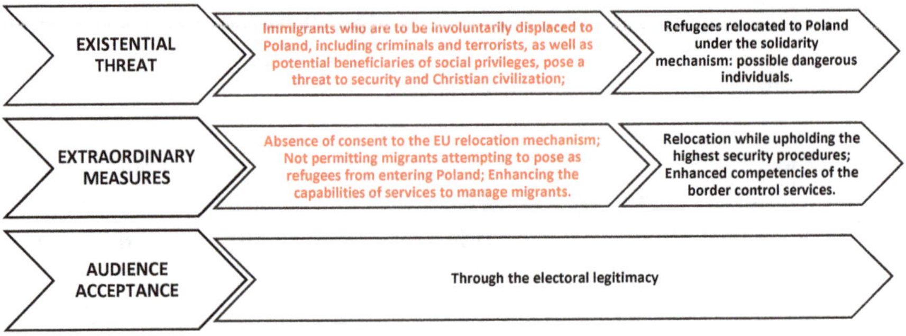

FIGURE 21 Securitization of migration issues in the narrative of the radical (red) and moderate (black) options
SOURCE: OWN STUDY

indirect, moderate stance: there is a threat associated with admitting refugees to Poland, but we can counteract it through procedures, inspections, checks, and strengthening services. A notable minority of debate participants took a humanitarian-solidarity approach. Figure 21 illustrates the process of securitization of migration issues among different groups of discourse participants.

The ongoing 2015 election campaign was replete with references to migration issues. The Left maintained the position preached in the debate: "Poles – victims of persecution and war – have been taken in by third countries many times in history. Therefore, Poland should accept those fleeing war, while taking care of their social integration and protecting public safety" (ZL, 2015). PO maintained the government's position, but in a softer form:

> We believe that European solidarity must be responsible, and we understand the principle of voluntarism as the ability to make a sovereign decision on the scale of our involvement. This must be accompanied by full government control over the process of accepting refugees, their smooth separation from illegally arrived economic migrants, the ability to vet newcomers for safety, and the provision of adequate funding in the EU budget. (PO, 2015)

PiS did not refer to the issue of migrants. In her exposé, Prime Minister Beata Szydło only stated: "The refugee issue makes us realize that the question of solidarity must be made clear. You can't call solidarity attempts to sort of export the problems that certain countries have created for themselves without the participation of others who are to be burdened with them" (Szydło 2015). There was more triumph in Prime Minister Mateusz Morawiecki's 2017 exposé: "The migrant relocation mechanism has not worked, other solutions are needed, and we want to take part in that discussion" (Morawiecki, 2017).

In the 2019 elections, only Zjednoczona Prawica (PiS) referred to the issue of migration, reminding us how successfully it prevented the relocation of migrants: Despite various pressures, not only did we not relent, but we caused a change in EU policy on this issue towards voluntarism (ZP, 2019). Mateusz Morawiecki summed up the government's policy in a similar way in his second exposé: "We convinced Europe of our solutions to the migrant issue, he triumphed, while boasting that: Today Poles are once again returning to their homeland. Finally today, Poland for generations of migrants can be a hope and a future, a place of return" (Morawiecki, 2019).

On July 7, 2021, a border crisis controlled by the Alexander Lukashenko regime began at the eastern border of the European Union with Belarus. This crisis involved the organized smuggling of migrants and refugees, primarily young men, into the EU from Iraq, Afghanistan, and other Middle Eastern and African countries. In response to EU sanctions imposed on Belarus following the rigged 2020 presidential elections, the Belarusian authorities launched and controlled these illegal smuggling channels. Consequently, the governments of Poland and Latvia decided to declare states of emergency in border areas with Belarus, while the Lithuanian government imposed a state of emergency nationwide. There was (and still is – in 2024) an overt instrumentalization of migrants by the Belarusian authorities to secure political concessions, such as the lifting of sanctions, and to provoke chaos. This unprecedented situation led to the construction of border barriers at the Union's eastern border and preceded another refugee crisis caused by Russian aggression against Ukraine in February 2022. At that time, approximately 4 million war refugees entered Poland within a few weeks, and after two years, around 950,000 remained (Eurostat, 2024).

Poland, like other countries of the European Union located in the eastern part of its territory, found itself in a new and unprecedented migration situation, which was reflected in the political discourse. The year 2022 was a series of 11 discussions and debates related to the presentation of *The Bill on Assistance to Citizens of Ukraine in Connection with the Armed Conflict on the Territory of This Country* and its changes (08.03.2022; 09.03.2022; 23.03.2022; 06.04.2022; 07.04.2022; 27.04.2022; 11.05.2022; 26.05.2022; 08.06.2022; 13.12.2022; 14.12.2022). The issues were securitized two-dimensionally – in the context of the security of Poland as a receiving country and in the context of the personal security of refugees. Representing the government side, Secretary of State at the Ministry of Internal Affairs and Administration Maciej Wąsik (PiS), who presented the assumptions of the law, said during its first reading:

> This law is a law about our solidarity towards Ukraine, a country that has suffered an unprecedented assault by Putin's Russia. This law is not

> only our moral duty, but it is also the Polish raison d'etat, because the Polish raison d'etat is a free, independent Ukraine. Today, when Ukraine is defended by men, soldiers, volunteers, we must take care of those who are fleeing from the horrors of war, fleeing because bombs are falling on their homes. These are mainly old men, women and children. It is our duty to make them feel safe in Poland. (Wasik, 2022)

This narrative, carried out, as it were, in an unspoken polemic to the 2015 migration debate, appealed not only to the needs of refugees who posed no threat to Poland, but also, more broadly, to building the country's security by supporting a struggling Ukraine. Hence the agreement to extraordinary measures, such as extending free medical care, education or social support to refugees without the need to work on Polish territory. Other major political groups present in parliament spoke in similar terms about: "passing the test of solidarity and openness" (Jacek Protas, PO), "a moral duty to the most defenseless victims" (Magdalena Biejat, Lewica), "unprecedented aid" (Krzysztof Gawkowski, Lewica), "taking care" (Bożena Żelazowska, KP) "of Ukrainian sisters and brothers" (Jarosław Gowin, Porozumienie), "solidarizing with the Ukrainian people unconditionally, unreservedly" (Pawel Rychlik, PiS), about the fact that "there are no refugee camps on Polish territory, because Ukrainians were taken in by Poles in their homes" (Magdalena Sroka, Porozumienie), "treating Ukrainian citizens equally with Polish citizens" (Jarosław Sachajko, Kukiz 15). This "made the refugees feel that here they can be taken care of, and this somehow allows them to look more optimistically into the future, to think that even if they cannot return to Ukraine, here they will feel safe" (Pawel Szramka, Polskie Sprawy). "A true Pole and a true patriot today is helping Ukraine and Ukrainians who are fighting for our freedom, who are fighting Putin today" (Piotr Borys, PO). Figure 22 shows the securitization of the issue of supporting refugees from Ukraine.

One party that was skeptical from the beginning and, as time went on, even opposed to supporting refugees was Konfederacja. Although it admitted, "the Polish nation shows great empathy and great every day, but it was afraid of treating Ukrainian citizens practically as citizens of the Polish state" and covering the cost of aid from the pockets of Polish taxpayers (Robert Winnicki). Representatives of the group talked about Poles, "who will have fewer rights than Ukrainians" (Konrad Berkowicz), "mass resettlement operations and identity threats" (Grzegorz Braun). Questions about the cost of aid and accusations were constantly coming from this politician: "You did not ask the Polish nation if it wishes to maintain the other nation (Braun). We cannot predict all the consequences of this massive migration. Therefore, out of prudence, we

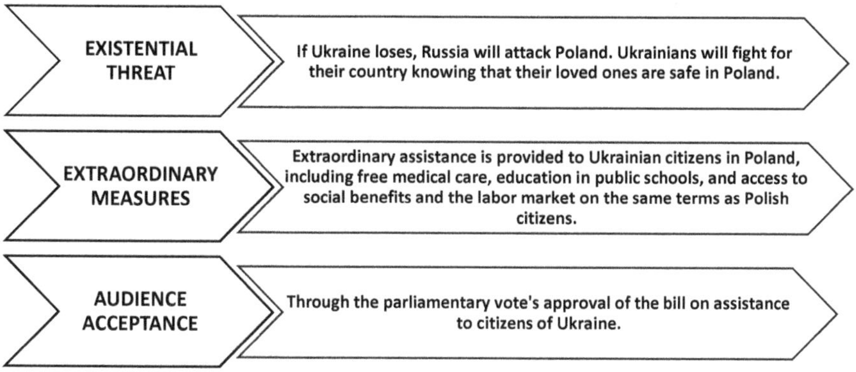

FIGURE 22 Securitization of the supporting refugees from Ukraine
SOURCE: OWN STUDY

should be concerned about precisely this asset, which is a group of people who have full civil rights in Poland" (Krzysztof Tuduj).

These concerns, as time went on, were also raised by groups advocating aid to Ukrainians: "Citizens are asking if we can afford it," reminded Magdalena Sroka (Porozumienie). "It must be remembered that very many Poles live in poverty. For while remembering the citizens of Ukraine, we must not forget that the state has first and foremost obligations to its own citizens," argued Joanna Senyszyn (PPS). "We will not accept any initiatives and amendments that aim to create better conditions for our guests from Ukraine than the citizens of the Republic of Poland have. Hospitality has its limits," said Jacek Protas (PO). The limits of support have been clearly defined, and their time scope has been limited.

The debate showed that most parties – though not without limits – support broadly assisting war refugees, keeping in mind both their personal security and Poland's security. Extraordinary solutions (the bill) were accepted, but only periodically. In 2023, the issue was desecuritized. In the election campaign, migration appeared only in relation to the crisis on the border with Belarus.

> "In 2021 Belarus carried out a hybrid attack on the Polish border," PiS reminded in its 2023 election program. – "Lukashenko brought migrants from Middle Eastern countries to the Polish-Belarusian border. The goal was to destabilize Poland and the European Union. Migration pressure, aggressive incidents on the border and provocations by Belarusian services forced the government to take swift action to secure the state border. A law on the construction of state border security was then prepared. The solution created the legal basis for rapid implementation of this

> investment. The physical security is 5.5 meters high. The system monitors a section of the border at a length of 206 km. The security consists of a system of cameras, cables and sensors." (PiS 2023)

"We will provide EU funding for the defense of the Polish border with Belarus. We will eliminate the route of smugglers passing from the Middle East through Belarus to Poland and on to the EU," PO echoed its rival. "My ambition is to lead a European policy on behalf of Poland to protect the border. Borders from all sides of Europe, so that the threat of illegal migration is not as acute as it is now," Donald Tusk (2023) declared in his expose.

The substantive scope of migration issues in Polish political discourse has changed over time. At the beginning of the period under study, it primarily focused on relations with the Polish diaspora scattered around the world, with its main thrust being support for Poles living abroad, especially in respecting their rights as national minorities and in supporting their repatriation aspirations. Over time, the category of others/aliens – viewed both in immigrants (especially illegal immigrants pretending to be refugees) and in selected national minorities (such as the German minority) who have lived in Poland for years – entered the discourse and began to gain increasing importance. These "others" may have been more or less culturally close, but they always received fewer rights and privileges than Poles. As the relevance of this subject increased, its securitization processes intensified. The third, less explored thread of the migration theme was the emigration of Poles in the 21st century to other EU countries that opened their labor markets to new EU members. This thread was also occasionally presented from the perspective of demographic security.

6 Environmental Security Discourse

The category of environmental security, referred to in the 1990s as ecological (the name environmental will appear in political discourse in the 21st century), already appeared in the first election programs in 1990. Politicians drew attention to the need to protect the environment and eliminate environmental threats. However, it took the form of general declarations, rarely supported by detailed proposals: "We consider saving the environment from progressive degradation to be one of our main tasks. We have a duty to do everything so that future generations are not left with a destroyed and poisoned – dangerous – country" (PPD, 1990). "We believe that the elimination of ecological threats is as important an issue in Poland today as economic development and the reconstruction of the political system" (PPS, 1990). "We regard environmental protection as one of the fundamental conditions for the successful development of society and the economy" (PSL, 1990, 1991). "Among the most urgent tasks now is the removal of threats to human health and life in areas of ecological disaster" (UD, 1991). "We will prevent ecological disaster through legal and systemic solutions that enforce safe methods of industrial and agricultural production" (NSZZ "S," 1993). "Ensure a noticeable improvement in the ecological security of the population" (PSL, 1997).

"We realize how bad the state of our environment is," Prime Minister Tadeusz Mazowiecki said. – "It is necessary to change economic strategy, spatial planning and design so that the environment we live in is not negatively affected," he postulated (Mazowiecki, 1989). "In the shadow of the crisis and reform of the economy was the problem of the drastic state of the environment in our country," Jan Krzysztof Bielecki (1991) said in a similar vein. "The government will make efforts to protect the environment," Jan Olszewski (1991) declared. "We will develop ecological education, which should be the responsibility of every citizen to realize environmental protection," Waldemar Pawlak announced (1993). "Proper environmental protection is one of the basic conditions for a dignified life," explained Jerzy Buzek (1997). The issue of environmental protection was considered important, its condition – a threat, but securitization processes did not follow. In the 2001 campaign, only PSL referred to the issue, making a general declaration: "We will ensure Poland's ecological security, guaranteeing better protection of the population's health and a good future for our country" (PSL, 2001). Prime Minister Leszek Miller spoke of a willingness to protect the environment and invest in this area (Miller, 2001), as did his successor, Marek Belka (2004). The issue did not appear in the 2005 campaign, but Prime Minister Kazimierz Marcinkiewicz had already referred to it: "the essential elements of our program are to ensure the ecological security of the country," he declared (2005).

In the 2007 campaign, the PSL again addressed the topic, listing environmental security in line with other categories, and by the Left, declaring that "measures are needed to achieve satisfactory national environmental security" (LiD, 2007). Later in the program, it considered environmental issues in the context of energy security. Representing PO Prime Minister Donald Tusk interpreted the state of the environment and investments in its improvement in the context of EU programs and funds. Bronislaw Komorowski (2010) spoke generally about respecting the environment.

The 2011 campaign addressed the issue in the programs of the three parties. PiS, referring to the category for the first time, drew attention to water shortages, advocating "the development and implementation of a water management strategy that considers both securing Poland against water shortages and flood security" (PiS, 2011). In the next paragraph, however, it noted that the challenges of environmentalism must take into account the needs of energy security provided largely by hard coal. SLD, like PiS, mentioned the environment and ecology in several places in the program, in passing in other categories. It accused "the fragmentation of services of combating environmental crime and the complete lack of state control over land use, which increases the risk of disasters such as floods, droughts, landslides, etc." (SLD, 2011). The PSL presented the broadest and multi-faceted program. At the outset, it was noted that the PSL considers "the rational combination of Poland's rapid development with the requirements of environmental protection as its duty and challenge." This declaration was followed by an enumeration of planned activities in the area of "environmental safety" (PSL, 2011).

Standing before parliament for the second time, Donald Tusk (2011) did not refer to the environment. His successor, Ewa Kopacz, declared:

> We realize the importance of environmental protection, to point out right away: The European Council summit will consider a draft of new targets for reducing gas emissions. My government will not agree to provisions that mean additional costs for our economy and higher energy prices. (Kopacz, 2014)

In this area, the PO spoke with one voice with PiS, prioritizing energy security over environmental concerns. In the 2015 campaign, the issue was raised again by three parties, but only one of them – the PSL – continued its earlier declarations. It repeated general assertions from previous years about linking development and ecology (PSL, 2015). Neither Beata Szydło's exposé (2015) nor Andrzej Duda's standing before the National Assembly (2015) addressed the topic. In contrast, Mateusz Morawiecki devoted three long paragraphs of his speech to the issue, focusing on the problem of air pollution: "48,000 Poles a year die prematurely because of smog, he alarmed, but immediately added: But we also realize that not everyone can afford cleaner fuel, which is why the program to fight smog is a program to support the poorest" (Morawiecki, 2017).

In the 2019 campaign, the issue was addressed by two electoral coalitions. KO declared "that it would lead to obtaining support from the EU budget for producing safe, cheap and cleaner energy, for programs to combat smog in Poland and for public transport" (KO, 2019). PiS dealt with asbestos and waste:

> State bodies responsible for environmental safety will act more intensively on these issues. (…) We have introduced a total ban on importing

> all types of waste for disposal and municipal waste into Poland. (…) We have increased penalties for violations of waste management regulations. We also introduced a fast track to remove "ecological bombs," i.e. abandoned hazardous waste. (ZP, 2019)

Both coalitions thus moved on the periphery of prevention, without addressing key issues such as the relationship between the environment and energy, the Green Deal being prepared by the European Commission, or even the problem of smog in Polish cities. The latter topic was addressed in his speech by Prime Minister Mateusz Morawiecki (2019), declaring "that Poles have the right for their children to breathe clean air." Later in the exposé, however, he saw no problems in the state of the Polish environment: "In many areas, we are greener than we think. About 3 million Poles drive gas-powered cars. We are developing an electromobility program."

The 2023 campaign was a record in terms of the amount of space devoted to environmental issues in the context of security. The most talked about in this regard was not by leftist or centrist groups, but by PiS. The environment-dedicated part of the party's program began with a description of the achievements of eight years in government:

> The government has carried out a comprehensive reform of environmental inspections. Periodic inspections of landfills, business entities and road transports are being carried out. Inspections work around the clock, laboratories and inspection equipment are modern and up to date with current environmental challenges. We have increased penalties for environmental offenses. We have introduced tough regulations to fight garbage mafias. The government's goal was to make Poles safer from emerging threats of waste dumping. (PiS, 2023)

The party claimed that it had improved water quality, increased the amount of greenery in cities, and protected forests from being bought up by foreigners. Significantly, the Green Deal and the relationship between concern for energy security and the state of the environment were not addressed. This, in turn, was the pivot of Konfederacja's program, which, speaking of

> climate ideology, maintained that, at the demand of the European Union, we are to deprive ourselves of energy security, and thus, in part, of political independence (…) A Poland without coal is a dangerous utopia. The EU's climate policy must be rejected and a separate energy transition path must be developed. Our priority in the EU must be to challenge the

> EU's Climate and Energy Package and negotiate a rebate in reducing carbon dioxide emitted by Poland. (Konfederacja, 2023)

Such a narrative, in which environmental security becomes a threat to energy security, and perhaps also economic and social security, has a good chance of becoming the basis for a new securitization in the future, given the rise of the zero-carbon issue and the resulting costs.

7 Conclusions

Discourse analysis in economic, societal, and environmental security reveals two examples in which the securitization of a single issue is carried out in different ways. This is the case with energy security, which is securitized differently by right-wing and center-left parties. The former view energy security in the context of geopolitical threats from countries that supply energy resources (particularly Russia), which may use them to exert pressure and undermine the sovereignty of recipient countries. Energy security is synonymous with economic independence, sovereignty, and even the foundation of autonomy. It is becoming one of the key elements of foreign policy. In contrast, for center-left groups, energy security is closely linked to energy transition – not only towards diversification but also a shift away from coal in favor of renewable energy sources. It also has an ecological dimension, understood as an aspect of ensuring environmental security. Finally, it is perceived not only in the Polish context of the Russian threat but more broadly as a result of global changes in energy markets.

The same applies to the issue of migration, where both the threat posed by uncontrolled migration from the MENA region, including the relocation of migrants already residing in other EU countries – a situation characteristic not only of Poland – and the issue of refugees from Ukraine who arrived in Poland after the Russian aggression of 2022 are securitized. Here, however, at least in the initial months of the event, securitization concerns the need to provide them with extraordinary assistance, as only such support to Ukraine will enable it to continue resisting Vladimir Putin's aggression effectively. These two different attitudes, presented simultaneously towards different groups of migrants, show that the securitization of one issue can have different faces and present a specific local context.

The securitization of measures to combat the SARS-CoV-2 pandemic proceeded in a manner typical of this process in other countries affected by the disease: reference was made to the uniqueness of the situation and the particular threat posed by a new, unknown virus, the need to take extraordinary measures such as increased funding for the healthcare system, restrictions on civil liberties and economic activity, and the requirement to behave responsibly towards other members of society. In turn, the securitization of radical changes to the pension system, necessary to provide pensions for an aging population, consisted of appealing to public acceptance of raising the retirement age – an issue that is hotly debated in Poland and meets with considerable resistance in society. This securitization was initiated but unsuccessful, as voters rejected the party that proposed this measure.

It may come as a surprise that the crucial issue of the economy's state after the political transformation of 1989 was not securitized, despite its significant impact on Polish households, which contributed to societal impoverishment in the early years following the political change. Although this issue featured in political discourse and was occasionally related to feelings of security, it remained unsecuritized: no existential threat was recognized, and no extraordinary measures were proposed to address it.

CHAPTER 6

Dynamics of Securitization in Polish Political Discourse

The noble ideas of the Fathers of Securitization Theory, which aimed to extend security research beyond the military realm while maintaining appropriate criteria to distinguish security from other types of policy (Gad and Petersen, 2011: 315), contributed to the observation and efforts to eliminate or reduce numerous significant threats to various aspects of human life. By drawing the attention of security actors to issues beyond military force and deterrence and convincing them that the general well-being of the individual is as crucial as mere survival, these concepts transformed the post-Cold War world. Despite the positive changes brought about by constructivists (and their critics), the instrumental use of theory for contemporary policy purposes was still prevalent. Although political discourse on security issues was often conducted substantively and in a balanced manner, instances arose where references to security were employed to shape perceptions of reality in accordance with the political objectives of the message sender. This frequently involved the manipulation of the political narrative, propaganda, and distortion, exacerbated by populism (Rittel, 2005: 306). This issue was also evident in Poland.

After 1989, Polish politicians recognized the significance of security issues in shaping their image and political standing. Lech Kaczyński's original approach to managing the Ministry of Justice from 2000 to 2001 focused primarily on responding to irregularities in the courts and prosecutor's offices reported by the media, particularly issues such as delays in proceedings and dismissals in criminal cases that outraged public opinion, as well as reduced sentences, which earned him the nickname of "sheriff" and contributed to his victory in the elections for the President of Warsaw. Bronisław Komorowski brought the experience gained as Minister of National Defense to the Presidential Palace, where he shaped the command system of the Polish Armed Forces and addressed other military leadership aspects, theoretically under the purview of the Ministry of Defense. Aleksander Kwaśniewski, by engaging the Polish Armed Forces in Iraq and Afghanistan, as well as supporting the Orange Revolution in Ukraine, anticipated elevated international positions in security after his presidency, with speculation about roles such as head of NATO or Secretary-General of the United Nations. Each of them, albeit in different forms, viewed the prioritization of security issues as a pathway to personal success in their political endeavors. Throughout the period under discussion (1989–2020), security issues remained a consistent and comprehensive component of political programs, strategies, debates, and politicians' speeches.

The main problem raised during the work on this book was the question about the importance of security issues and the dynamics of securitization processes in the Polish political discourse after 1989. Specific research problems related to the following:

1. Changes in the understanding of the concept of security and its threats over the last 30 years, reflected in political discourse;
2. Changes in the position of security issues in political discourse;
3. The level of securitization and riskification of the Polish political discourse;
4. Understanding and the manner of the use of security issues by different political parties and communities;
5. Fear and threatening as elements of political discourse;
6. Distinguishing between securitizations (extending the security category according to the Copenhagen School's assumptions), naming security issues without a securitization process, and tampering with the concept of security;
7. Desecuritization of selected areas previously recognized as exceptional issues due to belonging to the sphere of security.

The analysis of electoral programs, speeches by the Prime Ministers and Presidents, and selected parliamentary debates allowed the following conclusions to be drawn:

1. Security was an important element of Polish political discourse in the area analyzed in this book. In the 1990s, in Poland, as in other parts of the world, it was combined with the establishment of a military force capable of deterring and defending the country. This aim was pursued through both investment in the army and its qualitative transformation, as well as through political activities: first, to leave the Warsaw Pact peacefully and withdraw Soviet troops from Poland, and then to associate with and gain membership in NATO. While this issue was significant, the widespread agreement on the direction established in 1989 meant it did not require securitization. It was a security-related matter, but was implemented by the methods of "ordinary" policy.

   In the 21st century in Poland, as in other parts of the world, the category of security has expanded to include economic, social, and political issues. Periodically, issues such as the participation of the Polish Armed Forces in operations outside the country, ensuring energy security, cutting spending during the economic crisis, the perception of the migration crisis through the lens of Poland's internal security, or, as in 2020–2021, the health of Poles in the context of the COVID-19 pandemic became crucial and were securitized to add importance to them and garner support for their implementation and resolution. Political discourse here has followed global changes in the understanding of security and has responded to current threat scales.

2. Security issues occupied an important but relatively smaller space in political discourse throughout the analyzed period. On average, the issue covered around a dozen percent of every exposé of the Prime Ministers after 1989 (5.43% being the lowest and 34.04% the highest). In the case of Presidents, an upward trend can be observed (rising from 9.09% in Wojciech Jaruzelski's speeches to over 24% in both Andrzej Duda's speeches before the National Assembly). However, it should be noted that the initial speeches were of a different character than the contemporary ones: they were brief and conveyed the general view of the President being sworn into office, rather than outlining plans for specific areas.
3. Although security issues occupied a comparable "volume" of political discourse in the analyzed 30-year period, the robust intensification of securitization processes can be observed in the 21st century, after 2003 (Table 11, Figure 23).
4. Comparisons along the left-right spectrum regarding the amount of space devoted to security issues do not reveal a single trend. In the case of Prime Ministers, leaders from right-wing and center-right parties devoted an average of 15.9% of their exposés to security, while representatives of left-wing parties allocated 22.26%. However, in the case of Presidents, the situation is reversed: in the speeches of right-wing Presidents, security issues comprised an average of 20.45% of the programs (statistics were significantly understated by L. Wałęsa), whereas for left-wing Presidents, it was 9.96%.

Left-wing politicians in the post of Prime Minister refer to political security in international terms (relations with other countries, membership in organizations) more often than right-wing politicians do. In the case of Presidents, the situation is the opposite. However, a tendency for addressing issues related to social security cannot be observed among left-wing Presidents to a greater extent than among right-wing Presidents, which is considered to be the mark of the Left. The opposite is true – right-wing Prime Ministers and Presidents are more likely to address these issues. Right-wing politicians in both positions are also more likely to raise the issue of military security than their left-wing counterparts.

Therefore, it cannot be asserted that there were significant differences regarding the issue of security among Polish politicians from different parties in the domestic political discourse.

5. Security threats were used to instil fear in the audience of discourse. The most striking example is the narrative led by right-wing groups about the potential admission of migrants to Poland under the European relocation program. Possible visitors were portrayed as a diverse threat to Polish society (terrorists, spreaders of diseases, destroyers of Western culture and Christianity, and offensive Muslims). This narrative reflected the social mood at the time (in 2015–2016, 53–61% of respondents to the CBOS survey opposed receiving migrants) and enabled parties that threatened with the migration crisis to win parliamentary and presidential elections.

   The discourse of fear also emerged regarding the Iraqi issue – initially, it was claimed that not joining the coalition jeopardises Poland's security in a situation where it may require assistance in defense, and later that a prolonged presence there exposes not only the deployed soldiers but also the country to potential terrorist attacks.

   In other securitized issues, it was also warned what would happen in the absence of the proposed extraordinary measures; however, the intensity of this narrative was lower than in the examples mentioned above.

6. Simply calling something an issue of security does not constitute securitization. Politicians have repeatedly used this term to emphasize the issue, yet they often fail to propose extraordinary solutions to these security concerns. Sometimes, this is due to the absence of a need for such measures (the norms of ordinary political action suffice), while at other times it occurs because the process is merely rhetorical. The Securitization Theory does not always allow for the clear differentiation of these matters, but references to Just Securitization Theory – with its emphasis on assessing the intentions of the securitizing entity, the objectivity of the threat, and the actions taken – more precisely highlight the difference between genuine securitization and manipulation or compelling rhetoric for the audience. Therefore, the issues outlined in the table above, selected from the broad spectrum of security topics in Polish public discourse, were deemed to be subjected to the securitization process.

7. Desecuritization is a process observed in the analyzed examples of discourse. Sometimes its phase is visible and ends with the uniqueness of the securitized issue, transferring it to the rules of ordinary policy (an example being the collapse of the pension system). At other times, it

TABLE 11 Selected issues of Polish political discourse in light of the Classical Securitization Theory, Just Securitization Theory and Riskification Theory

The examined issue	Classical Securitization Theory Copenhagen School (B. Buzan, O. Waever, O. de Wilde)			
	Existential threat	Extraordinary measures	Approval of the public	Desecuritization
Crime (multi-dimensional: organized vs "ordinary," economic) 2001–2015	It is a threat to a citizen who is afraid to leave his/her home or leave his/her car in the parking lot. Even the chief police officer was killed.	Tougher penalties, streamlining procedures, absolute and rapid response, zero tolerance	Partial, in election by the support of the party which proposed it	Yes, the issue moves into the area of riskification
Participation of the Polish Armed Forces in the operation in Iraq 2003–2007	WMD + terrorism, human rights violations	Participation of the Polish Armed Forces in the overthrowing of the regime, and then in the stabilization operation alongside the allies	Partial of the population in general (according to opinion polls), majority in the case of political elites	Yes, at the time of the decision to withdraw the troops
Corruption 2005–2015	Corruption causes huge losses for the budget and demoralizes the political class, harming the ordinary citizen, supports organized crime	Extraordinary penalties, property statements and its verification systems, establishment of a special anti-corruption unit	Partial, in election by the support of the party which proposed it	Yes, the issue moves into the area of riskification

Just Securitization Theory (R. Floyd)				Riskification Theory (O. Corry)		
Intentions of the securitizing actor	**Objectivity of a threat**	**Safety practice**	**Desecuritization/ resecuritization**	**Hazard risk**	**Emergency preventive actions**	**Approval of the public**
It is difficult to assess the intentions of the securitizing entity. Securitization was part of the electoral strategy.	The threat was serious, confirmed by statistics.	Most of the proposed solutions were introduced, The Central Bureau of Investigation (Police) was established	Yes, the issue moves into the area of riskification	New challenges in the fight against crime (especially in cyberspace)	Lack of consent to the liberalization of the penal code, continuous requirement to improve the effectiveness of prosecutions and speed of proceedings	Yes
The intentions were different from the actual reasons – overthrowing a tyrant and fighting weapons of mass destruction and terrorism – it was about building an allied credibility.	It is difficult to indicate a direct link with S. Hussein's activities and the security of the Republic of Poland	It happened – the Armed Forces took part in the overthrowing of the regime and then the stabilization operation.	Yes, at the time of the decision to withdraw the troops	The instability of the Iraqi state, the strengthening of terrorist groups operating on its territory, the risk of oil price fluctuations	Consulting and training operations	Implied, due to the low risk of the operation
It is difficult to assess the intentions of the securitization entity. Securitization was part of the electoral strategy.	The threat was serious, confirmed in statistics and reports (e.g. Transparency International)	Most of the proposed solutions were introduced, the Central Anti-Corruption Bureau was established.	Yes, the issue moves into the area of riskification	The dangers are not eliminated, still a lot of attention must be paid to it.	Continuous improvement of procedures and stricter requirements for public figures	Probably yes

TABLE 11 Selected issues of Polish political discourse in light of the Classical Securitization (*cont.*)

The examined issue	Classical Securitization Theory Copenhagen School (B. Buzan, O. Waever, O. de Wilde)			
	Existential threat	**Extraordinary measures**	**Approval of the public**	**Desecuritization**
Poland's energy security 2006–2019	The supply of energy resources to Poland, especially from Russia, is at risk. It will be like in Ukraine, where gas supply was interrupted in 2006.	It is necessary to stop the North Stream, diversify energy sources (construction of LNG terminal, other than Russian suppliers). NATO needs to be involved in energy security and the EU must oppose the Northern Stream.	Partial, in election by the support of the party which proposed it	Yes, after most of the planned actions have been completed
NATO/UN/EU foreign operations, including Afghanistan 2007–2019				
Economic crisis 2011	Poland, like Europe, is experiencing economic slowdown	Austerity plan, cuts in social spending	Partial (public opinion polls)	Yes, at the end of the term of office of the government of the parties that securitized the issue.

	Just Securitization Theory (R. Floyd)				Riskification Theory (O. Corry)		
Intentions of the securitizing actor	**Objectivity of a threat**	**Safety practice**	**Desecuritization/ resecuritization**	**Hazard risk**	**Emergency preventive actions**	**Approval of the public**	
Intentions can be considered fair	The threat is objective – Russia pursues its policy with the help of energy resources	It happened – the LNG terminal was built in Świnoujście, contracts for gas from the USA and Norway, investments in renewable energy sources	Yes, after most of the planned actions have been completed	While much has been done to eliminate the threat, there is still a risk of instability in the supply of energy raw materials to Poland.	Further diversification, even more renewable energy, related to energy security energy, gradually withdrawing from coal	Yes (public opinion polls)	
				Terrorism, WMD, human rights violations	Military operations, but rapidly moving into stabilization, advisory and training operations. Permanent but qualitatively limited presence in conflict areas as a guarantee to halt potential escalation	Yes, after being presented as an alternative to intensive military action	
Intentions can be considered fair	The economic crisis was a fact – the threat was objective, but it was possible to discuss its scale.	Partial implementation of plans interrupted by elections and change of approach to the problem	Yes, due to the end of the term of office				

TABLE 11 Selected issues of Polish political discourse in light of the Classical Securitization (*cont.*)

The examined issue	Classical Securitization Theory Copenhagen School (B. Buzan, O. Waever, O. de Wilde)			
	Existential threat	**Extraordinary measures**	**Approval of the public**	**Desecuritization**
Collapse of the pension system 2011	Pensions of currently working Poles will be dramatically low, the system will not be able to finance them	Extending the retirement age, taking over some of the funds accumulated in OPFs by ZUS to finance the system	Partial (public opinion polls). The next election was a disaster for the parties that made it happen.	Yes, at the end of the term of office of the government of the parties that securitized the issue.
Defending the territory against the military threat from Russia 2014–2020	After Russia's expansion to Georgia and Ukraine, Poland is another natural destination. The attack from this side is only a matter of time.	It is necessary to lead to the presence of NATO (especially US) troops in the territory of the Republic of Poland, increase expenditure on the military, establish Territorial Defense Forces	Partial, in election by the support of the party which proposed it	No
Migration crisis in Europe 2014–2019	The influx of terrorists claiming to be refugees into Poland. Cultural, health and criminal hazard.	The EU mechanism for relocating refugees to countries with fewer refugees, including Poland. Should be blocked	Partial, in election by the support of the party which proposed it	No, but the issue has a chance to move into a riskification phase.
Coronavirus pandemic 2020	Death and permanent damage to the health of many people, the collapse of the health care system resulting in the cessation of treatment of other diseases as well	New procedures for staying in public space, lockdown, increased expenditure on health care system	Partial (public opinion polls). Despite the restrictions introduced, the next election was won by the party that was responsible for it.	No visible moment of desecuritization, transition to a riskification phase

SOURCE: OWN DEVELOPMENT

Just Securitization Theory (R. Floyd)				Riskification Theory (O. Corry)		
Intentions of the securitizing actor	**Objectivity of a threat**	**Safety practice**	**Desecuritization/ resecuritization**	**Hazard risk**	**Emergency preventive actions**	**Approval of the public**
Intentions can be considered fair	All economic forecasts and analyses predicted a collapse of the pension system in the absence of reforms	Partial implementation of plans interrupted by elections and change of approach to the problem	Yes, due to the end of the term of office			
Intentions can be considered fair	The threat posed by Russia seems to be real and huge, but not in classical military terms, as it is presented to the public.	The demands were met – NATO troops went to Poland, increased spending on the military to 2% GDP, Territorial Defense Forces were established	No			
The main intention of the securitization entity was to obtain greater electoral support.	The scale of the threat, given the number of potential refugees to arrive in Poland, was small.	Relocation has been halted.	No, but the issue has a chance to move into a riskification phase.			
Intentions can be considered fair	The threat was real, but not always as huge as the proposed measures to combat it.	Fulfillment of all demands	No visible moment of desecuritization, transition to a riskification phase	The virus will stay with us for long forever	Vaccination plus maintaining hygienic safety rules	Partial (public opinion poll plus high interest in the vaccination program)

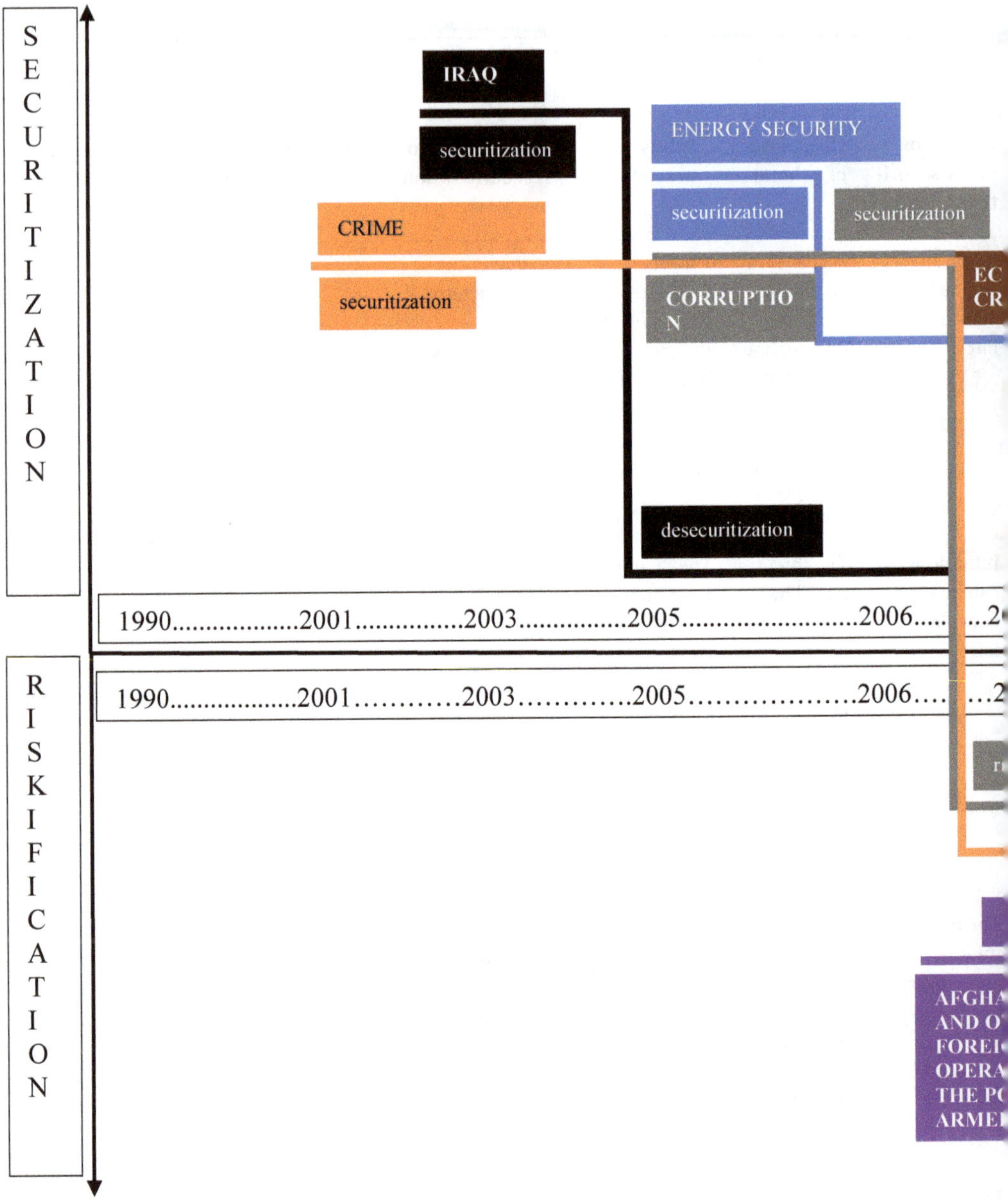

FIGURE 23 Selected issues of the Polish political discourse in light of the Classical Securitization Theory, Just Securitization Theory and Riskification Theory
SOURCE: OWN STUDY

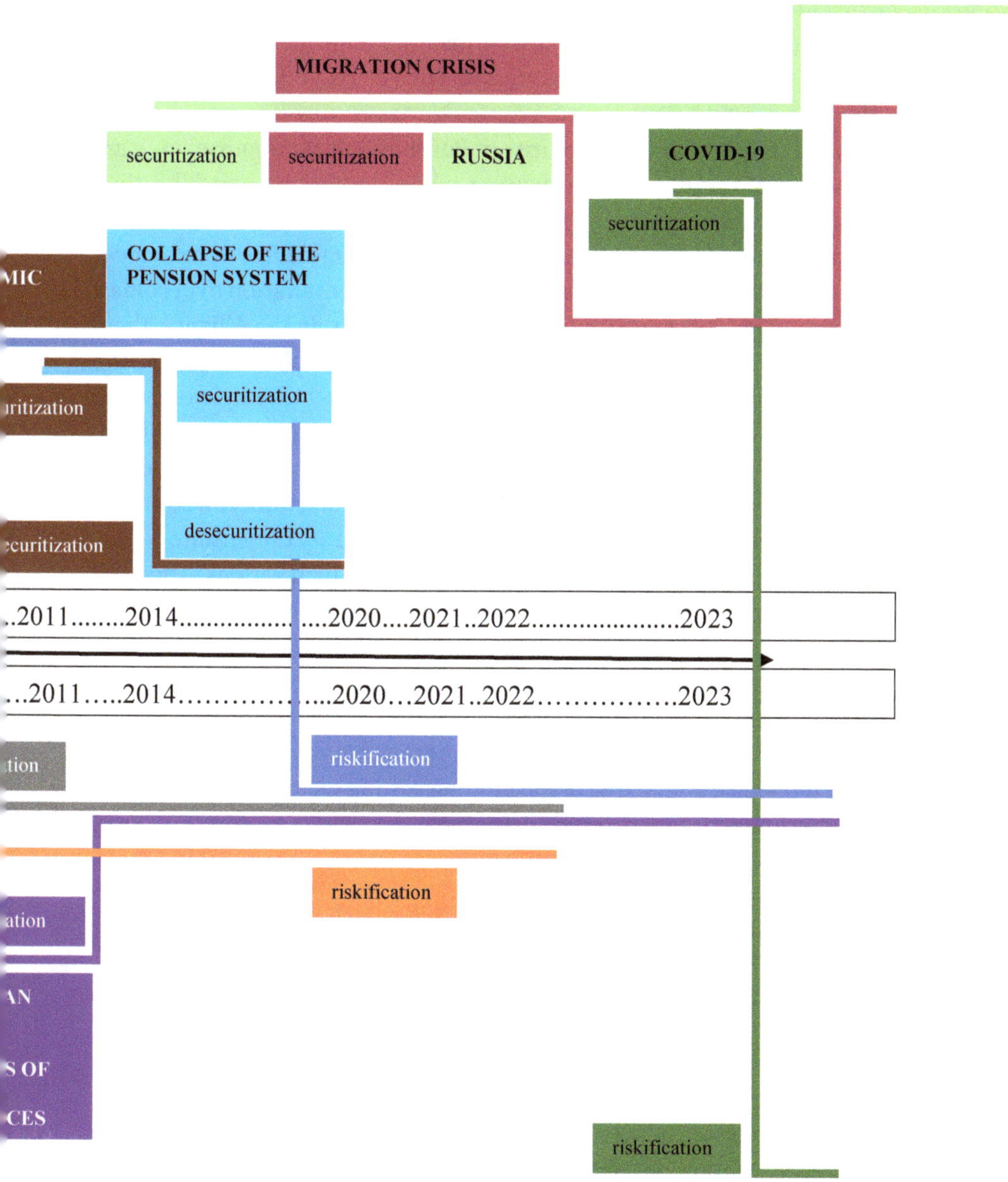
MIGRATION CRISIS
securitization
securitization
RUSSIA
COVID-19
securitization
MIC
COLLAPSE OF THE PENSION SYSTEM
iritization
securitization
:curitization
desecuritization
..2011........2014...................2020....2021..2022.....................2023
...2011.....2014...................2020...2021..2022...............2023
tion
riskification
riskification
ation
AN
S OF
CES
riskification

coincides with the transfer of the issue to the riskification phase. It seems, however, that sooner or later, desecuritization is inevitable, even in cases where issues are still being securitized, such as migration problems and the threat from Russia.

Out of the 18[1] selected areas of security in Polish political discourse, nine are cases related to a securitization mechanism, while one pertains to riskification. The remaining nine issues are referred to as security concerns, but there is no securitization process (no proposals for extraordinary measures or references to public approval). In some instances, such as Poland's aspiration to become a NATO member and subsequently building a position within the Alliance, there was no need to propose extraordinary actions or seek public opinion, as the issue was recognized as important and acted upon under "ordinary" policy. The chosen direction was not questioned by significant public figures, and there was widespread understanding of the need to join a robust transatlantic alliance, which was ultimately achieved. A similar mechanism occurred in the context of European integration, particularly in its security dimension.

The analyzed discourse also includes situations in which various aspects of social, economic, health, or environmental security are mentioned; however, this is a rhetorical procedure, branding a certain area as a matter of security without transitioning to a securitization act. These initiated and unfinished securitizations often occur with merely the use of the term, lacking any attempt to develop the outlined themes or present them from a security perspective. Frequently, after a title or heading suggesting sectoral security issues, the topic is examined from entirely different perspectives (e.g., the narrative in the electoral programs of the Polish People's Party (PSL) from the 21st century). Therefore, in quantitative discourse analyses, one should be cautious regarding the use of the term security without conducting a qualitative analysis of the context in which it appears, as this may lead to erroneous conclusions. The meta-operational role of the concept of security, along with the universality

1 (1) Poland's membership in NATO and the country's position in the Alliance; (2) strategic partnership with the USA; (3) The European Union as a security guarantor; (4) The Russian Federation as a military threat; (5) the condition of the Armed Forces, their modernization and financing military spending; (6) the threat of a conventional attack on Poland (territorial infringement); (7) involvement in the operation in Iraq; (8) involvement in the operation in Afghanistan and other foreign operations of the Polish Arms; (9) anti-crime policy; (10) corruption and the fight against corruption; (11) economic security of the state; (12) energy security; (13) social security, pension reform; (14) the migration crisis, (15) the economic crisis; (16) production of food in order to ensure national security; (17) environment as a matter of security; (18) COVID-19 pandemic.

of references lacking deeper content, can create a misleading impression that everything is security, which is factually inaccurate.

In nine separate cases of securitization (Table 11), it is noteworthy that most of these situations also underwent a desecuritization process, after which they were either no longer regarded as a security issue or transitioned into the riskification phase. This process is particularly evident in the case of the Polish Armed Forces' involvement in the operation in Iraq. The issue was heavily securitized at the time the decision was made to send the Polish Armed Forces to confront Saddam and then stabilize Iraq; however, once such operations became standard in NATO policy, it was reclassified as "ordinary" policy (it was desecuritized). Numerous military engagements in Afghanistan and other foreign operations involving Polish soldiers in the 21st century followed the logic of riskification: Poland participates in these projects due to the risk that issues from other parts of the world may also impact Europe.

The transition from securitization to riskification is a common phenomenon in Polish political security discourse. This is especially evident in the case of energy issues, which were strongly securitized during periods of low diversification in raw material supplies. After several years and an increased number of sources for raw materials, these issues entered a riskification phase. As Prime Minister M. Morawiecki pointed out in the speech quoted in Chapter 4, there is no imminent threat of supply interruptions; however, there remains a risk that this process may be disrupted. A similar situation is evident with the COVID-19 pandemic: it transitioned from strong securitization in the spring of 2020 to a riskification phase characterized by months of securitization focused on minimizing the risk of further waves, increased disease cases, the emergence of virus mutations, and so on. As Corry predicted (2012), drawing on examples from health and environmental security, the trend of responding with extraordinary measures (although nominally less intense than securitization) is becoming a significant aspect of shaping long-term security policies. While it is true that two securitized issues – the migration crisis and Russia as a military threat – remain in the process of securitization, they exhibit a volatile and lower intensity than at the outset, showing a tendency to transition into a riskification phase.

The situation described in this chapter raises two problems on which both securitization actors and entities, as well as the public, will have to take a stand. First, since the securitization processes intensified in the 21st century, is there an increased risk that they are conducted not out of pure intentions to solve the problem, but due to political cynicism aimed at gaining and maintaining power under extraordinary conditions? Can violations of the law, including human rights, be justified by the promise of security? How can this be reconciled, especially in Europe, with the concept of an open society? Isn't it necessary, as Floyd (2019) implies, to consider the moral implications of expanding the category of security to prevent it from being overwhelmed by political populism and aggressive discourse?

And secondly, since security issues under extraordinary conditions increasingly often include not objective, existential threats but risks that may occur in the future (riskification), which indefinitely prolongs emergency states, procedures, and security practices, will everything not become security in a moment, and security itself become nothing? Should we be more concerned about the tyranny of security procedures or rather the lack of elimination of real threats?

Conclusions

Security is not merely a means of battling the enemy and addressing fears; rather, it often creates them, Jef Huysmans (2014: 3) noted, emphasising the significance of the processes of securitization and the broadening of the security category. According to Securitization Theory, threats do not exist consistently, in the same dimension and intensity, somewhere in the universe, waiting to be identified and addressed. The formation of security is a political process in which issues are transformed into security threats through a series of events framed within a securitization cycle. At times, all the conditions of the theory are met, while other times, there is merely a speech act that labels an issue as a security problem (which can sometimes be distorted by the populist behavior of the securitizing actor). Additionally, there are areas such as migration or environmental degradation, where inclusion in the category occurs not because they currently pose a real threat, but due to their potential to escalate to that level, affecting multiple sectors of security (economic, cultural, political, people, etc.), and are thereby riskified.

Securitization is a politically convenient procedure. It allows raising the relevance of a problem by presenting it as a security issue, thanks to the speech act so readily practiced in politics. The intensity of the use of this opportunity in the 21st century, also evident in Polish political discourse, coupled with the emergence of riskification processes, raises the question of the values or resources that one seeks to protect by extraordinary means. As mentioned at the beginning of this book, securitization processes have done much good for marginalized but important problems of the modern world (environmental threats, HIV), drawing attention to the fact that human security goes beyond tanks, multi-role aircraft, and diplomatic actions. However, there is also the other side of the coin. The potentially widespread negative consequences of securitization mean that its moral acceptability is already in everyone's interest (Floyd, 2019: 23).

It should not be forgotten that securitization holds significant power: it leads to political decisions that impact the lives of societies. Depending on the circumstances and the subject of securitization, it can result in the banning of certain industries whose production threatens the environment; it can authorize the forcible detention of people considered carriers of infectious diseases; it can initiate the construction of physical borders (walls and fences) and even the potential for shooting at individuals to stop the influx of illegal migrants; and it can be responsible for granting additional powers to the police and military or expanding a wide range of actions to combat the terrorist threat (Herington, 2012). This is why examining the intentions of securitising actors seeking to expand the security category is so important. When agreeing to such circumstances, it is essential to assess whether the threat is genuinely existential in relation to the reference object that it aims to protect, and whether the same effect can be achieved through "ordinary" policies, using standard solutions and measures. In making this postulation, Floyd cites the example of the wide range of extraordinary measures taken by Western countries, including the US in particular, in the post-2001 fight against terrorism, which she believes were excessive in relation to the reality of the threat. Similar concerns are being raised about measures taken around the world to combat the Sars-CoV-2 pandemic (Shani, 2020; Bilmes, 2020; Butler, 2020; Takao, 2020; Nurhasanah et al., 2020; Hastoro and Sumarni, 2020; Albert et al., 2021; Abbott, 2021; Villarreal, 2021). Researchers are starting to highlight the human security constraints caused by restrictive lockdown policies, social isolation, or pressure to vaccinate against the virus, indicating that health security must not overshadow the other needs and rights of individuals and entire societies.

The question of the morality of securitization processes is an urgent issue. Their intensity in the 21st century may raise suspicions about the motives behind them and the objective existentiality of the threats they address. The effectiveness of securitization processes presents an additional temptation here. However, the aim is not to move away from expanding categories of security, providing a new perspective on phenomena not previously regarded as security issues. While they may be viewed as such, extraordinary measures are not always necessary to mitigate threats, especially if they are not existential in nature. The question of riskification processes also remains open. It is indisputable that the world is increasingly focusing on risks, wisely reasoning that their elimination, or at least control, can prevent risks from materialising. However, excessive reliance on extraordinary measures at the risk stage can exacerbate the dangerous processes highlighted by Floyd (2019). Safety etiquette cannot permit every action under all circumstances. Extraordinary measures should be reserved for emergency situations that cannot be resolved by normal policy. Otherwise, security will lose its effectiveness and undermine the process of achieving a state free from significant threats.

Appendix

Polish Political Party Map

TABLE A1 Elections, coalitions, governments, prime ministers and presidents in the Republic of Poland: 1989–2023

Elections	Prime Minister	Dates	PM's party	Coalition	President
1989: June 4th and 18th	Tadeusz Mazowiecki	20.08.1989–04.01.1991	KO "S"	UD, PSL, SD, ROAD, FPD okresowo: PZPR, ZSL, SD	Wojciech Jaruzelski (1989–1990)
	Jan Krzysztof Bielecki	04.01.1991–06.12.1991	KLD	KLD, PC, ZChN, SD, ROAD, UD	
1991: October 27th	Jan Olszewski	06.12.1991–05.06.1992	PC	PC, ZChN, PSL-PL, SLCh	Lech Wałęsa (1990–1995)
	Waldemar Pawlak	05.06.1992–10.07.1992	PSL	Government wasn't formed	
	Hanna Suchocka	10.07.1992–26.10.1993	UD	ZChN, UD, KLD, SLCh, PSL-PL, PChD, PPPP	
	Waldemar Pawlak	26.10.1993–07.03.1995	PSL	SLD, PSL	
1993: September 19th	Józef Oleksy	07.03.1995–07.02.1996	SLD	SLD, PSL	
	Włodzimierz Cimoszewicz	07.02.1996–31.10.1997	SLD	SLD, PSL	Aleksander Kwaśniewski (1995–2005)
	Jerzy Buzek	31.10.1997–19.10.2001	AWS	AWS, UW (temporary)	
1997: September 21st	Leszek Miller	19.10.2001–02.05.2004	SLD	SLD, UP, PSL	
2001: September 23rd	Marek Belka	02.05.2004–31.10.2005	SLD	SLD, UP	
	Kazimierz Marcinkiewicz	31.10.2005–14.07.2006	PiS	PiS, Samoobrona RP, LPR	
	Jarosław Kaczyński	14.07.2006–16.11.2007	PiS	PiS, Samoobrona RP, LPR	Lech Kaczyński (2005–2010)
2005: September 25th	Donald Tusk	16.11.2007–22.09.2014	PO	PO, PSL	

TABLE A1 Elections, coalitions, governments, prime ministers and presidents (*cont.*)

Elections	Prime Minister	Dates	PM's party	Coalition	President
2007: October 21st	Ewa Kopacz	22.09.2014–16.11.2015	PO	PO, PSL	Bronisław Komorowski (2010–2015)
2011: October 9th	Beata Szydło	16.11.2015–11.12.2017	PiS	PiS, SP, Porozumienie	
2015: October 25th	Mateusz Morawiecki	11.12.2017–13.12.2023	PiS	PiS, SP, Porozumienie	Andrzej Duda (2015–...)
2019: October 13th 2023: October 15th	Donald Tusk	13.12.2023–...	KO	KO, Polska 2050, PSL, Nowoczesna	

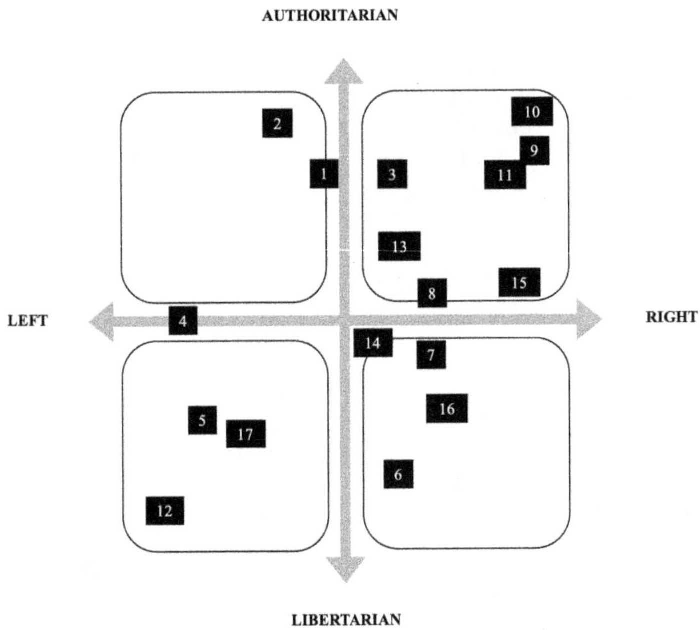

FIGURE A1 Positioning of Polish political parties based on Chapel Hill Expert Survey trend file (Seth et al., 2022)

TABLE A2 Polish political parties

No.	Polish name	Acronym	Previously
1	Prawo i Sprawiedliwość	PiS	Porozumienie Centrum (PC)
2	Solidarna Polska	SP	
3	Porozumienie		
4	Sojusz Lewicy Demokratycznej	SLD	Socjaldemokracja RP (SdRP) Lewica Lewica i Demokraci Zjednoczona Lewica
5	Razem		
6	Nowoczesna		
7	Koalicja Obywatelska	KO	Platforma Obywatelska (PO)
8	Polskie Stronnictwo Ludowe	PSL	Unia Polityki Realnej (UPR) – Partia Korwin
9	Partia Wolność		
10	Konfederacja		
11	Kukiz 15		
12	Ruch Palikota		
13	Samoobrona		
14	Unia Wolności	UW	Unia Demokratyczna (UD)
15	Akcja Wyborcza Solidarność	AWS	
16	Polska 2025		
17	Kongres Liberalno-Demokratyczny	KLD	

SOURCE: OWN STUDY

References

Abbott, L. (2021, January 20). Global Covid-19 Responses Through a Critical Security Studies Perspective. *E-International Relations*. https://www.e-ir.info/2021/01/20/global-covid-19-responses-through-a-critical-security-studies-perspective/.

Ädel, A. (2010). How to use corpus linguistics in the study of political discourse. In A. O'Keeffe & M. McCarthy (Eds.), *The Routledge Handbook of Corpus Linguistics* (pp. 591–604). Routledge.

Alker, H. R. (2005). Emancipation in the Critical Security Studies Project. In K. Booth (Ed.), *Critical Security Studies and World Politics* (pp. 189–213). Lynne Rienner Publishers.

Allenby, B. R. (2000). Environmental security: Concept and implementation. *International Political Science Review, 21*(1), 5–21. https://www.jstor.org/stable/1601426.

Amin-Khan, T. (2012). New Orientalism, securitisation and the western media's incendiary racism. *Third World Quarterly, 33*(9), 1595–1610. https://doi.org/10.1080/01436597.2012.720831.

Antoszewski, A. (1998). Proces i zmiana polityczna. In A. W. Jabłoński & L. Sobkowiak (Eds.), *Studia z teorii polityki* (189–203). Wydawnictwo Uniwersytetu Wrocławskiego.

Aradau, C. (2004). Security and the democratic scene: Desecuritization and emancipation. *Journal of International Relations and Development, 7*(4), 388–413. https://doi.org/10.1057/palgrave.jird.1800030.

Aradau, C. (2006). Limits of security, limits of politics? A response. *Journal of International Relations and Development*, 9(1), 81–90. https://doi.org/10.1057/palgrave.jird.1800073.

Austin, J. L. (1993). *Mówienie i poznawanie. rozprawy i wykłady filozoficzne*. Wydawnictwo Naukowe PWN.

Awdiejew, A. (2004). Systemowe środki perswazji. In P. Krzyżanowski & P. Nowak (Eds.), *Manipulacja w języku. Manipulacja w tekście* (pp. 71–79). Perswazja.

Awdiejew, A. (2007). *Gramatyka interakcji werbalnej*. Wydawnictwo Uniwersytetu Jagiellońskiego.

Baldwin, D. A. (1995). Security Studies and the End of the Cold War. *World Politics, 48*(1), 117–141.

Baldwin, D. A. (1997). The Concept of Security. *Review of International Studies, 23*, 5–26. doi:10.1017/S0260210597000053.

Balzacq, T. (2005). The three faces of securitization: Political, agency, audience, and context. *European Journal of International Relations, 11*(2), 171–201. https://doi.org/10.1177/1354066105052960.

Balzacq, T. (Ed.). (2011). *Securitization theory. How security problems emerge and dissolve*. Routledge.

Beaulac, S. (2004). The Westphalian Model in Defining International Law: Challenging the Myth. *Australian Journal of Legal History, 8*, 181–213. https://ssrn.com/abstract=672241.

Beck, U. (2012). *Społeczeństwo światowego ryzyka. W poszukiwaniu utraconego bezpieczeństwa*. Wydawnictwo Naukowe PWN.

Bellamy, A. J. (2009). *Responsibility to protect: the global effort to end mass atrocities.* Polity Press.

Bellany, I. (1981). Towards a Theory of International Security. *Political Studies, 29*(1), 100–105. https://doi.org/10.1111/j.1467-9248.1981.tb01276.x.

Bethurem, N. L. (2002). Environmental destruction in the name of national security: Will the old paradigm return in the wake of September 11? *Hastings West-Northwest Journal of Environmental Law & Policy, 8*, 109–127. https://repository.uchastings.edu/hastings_environmental_law_journal/vol8/iss2/1.

Bigo, D. (2002). Security and Immigration: Toward a Critique of the Governmentality of Unease. *Alternatives: Global, Local, Political, 27*, 63–92. http://www.jstor.org/stable/45468068.

Bigo, D. (2006). Internal and External Aspects of Security. *European Security, 15*(4), 385–404. https://doi.org/10.1080/09662830701305831.

Bigo, D. (2012). Międzynarodowa socjologia polityczna. In P. D. Williams (Ed.), *Studia bezpieczeństwa* (pp. 113–127). Wydawnictwo Uniwersytetu Jagiellońskiego.

Bilgin, P. (1999). Security studies: Theory/practice. *Cambridge Review of International Affairs, 12*(2), 31–42.

Bilgin, P. (2010). The 'Western-centrism' of security studies: 'Blind spot' or constitutive practice? *Security Dialogue, 41*(6), 615–622. https://doi.org/10.1177/0967010610388208.

Bilgin, P. (2011). The politics of studying securitization? The Copenhagen School in Turkey. *Security Dialogue, 42*(45), 399–412. https://doi.org/10.1177/0967010611418711.

Bilgin, P. & Morton, A. D. (2002). Historicising representations of 'failed states': Beyond the cold-war annexation of the social sciences? *Third World Quarterly, 23*(1), 55–80. https://doi.org/10.1080/01436590220108172.

Bilmes, L. J. (2020). Rethinking U.S. National Security after Covid19. *Peace Economics, Peace Science and Public Policy, 26*(3), 20200055. https://doi.org/10.1515/peps-2020-0055.

Booth, K. (1991). Security and Emancipation. *Review of International Studies, 17*(4), 313–326. https://doi.org/10.1017/S0260210500112033.

Booth, K. (1997). Security and Self: Reflections of a Fallen Realist. In K. Krause & M. C. Williams (Eds.), *Critical Security Studies: Concepts and Cases* (pp. 83–120). UCL Press. https://doi.org/10.1177/00108369921961889.

Booth, K. (2005). *Critical Security Studies and World Politics*. Lynne Rienner Publishers.

Booth, K. (2007). *Theory of World Security*. Cambridge University Press.

Bourdieu, P. (2005). *Dystynkcja. Społeczna krytyka władzy sądzenia*. Scholar.

Bourne, M. (2014). *Understanding security*. Palgrave Macmillan.

Budge, I. (1994). A new spatial theory of party competition. *British Journal of Political Science*, *24*(4), 443–467. https://doi.org/10.1017/S0007123400006955.

Budge, I., Klingemann, H. D., Volkens, A., Bara, J., & Tanenbaum, E. (2001). *Mapping Policy Preferences: Estimates for Parties, Electors, and Governments 1945–1998*. Oxford University Press.

Budge, I., Klingemann, H. D., Volkens, A., Bara, J., & McDonald, M. (2006). *Mapping Policy Preferences: Estimates for Parties, Electors, and Governments in Eastern Europe, European Union and OECD 1990–2003*. Oxford University Press.

Butler, C. (2020). Plagues, Pandemics, Health Security, and the War on Nature. *Journal Of Human Security*, *16*(1), 53–57. doi:10.12924/johs2020.16010053.

Buzan, B. (1983). *People, states, and fear: The national security problem in international relations*. Wheatsheaf Books.

Buzan, B. (1991). *People, states, and fear: an agenda for international security studies in the post-Cold War era*. Harvester Wheatsheaf.

Buzan, B., & Hansen, L. (2009). *The Evolution of International Security Studies*. Cambridge University Press. https://doi.org/10.1017/CBO9780511817762.

Buzan, B., Wæver, O., & de Wilde, J. (1998). *Security: A new framework for analysis*. Lynne Rienner Publisher.

Buzan, B. & Wæver, O. (2003). *Regions and Powers: The Structure of International Security*. Cambridge University Press.

Caballero-Anthony, M., Emmers, R. & Acharya, A. (2006). (Eds.). *Non-traditional security in Asia: dilemmas in securitization*. Ashgate Publishing.

Campbell, D. (1992). *Writing Security. United States Foreign Policy and the Politics of Identity*. University of Minnesota Press.

Christou, O. & Adamides, C. (2013). Energy securitization and desecuritization in the New Middle East. *Security Dialogue*, *44*(5–6), 507–522. https://doi.org/10.1177/0967010613499786.

Ciută, F. (2010). Conceptual notes on energy security: Total or banal security? *Security Dialogue*, *41*(2), 123–144. https://doi.org/10.1177/0967010610361596.

Corry, O. (2012). Securitisation and 'Riskification': Second-order Security and the Politics of Climate Change. *Millennium: Journal of International Studies*, 40(2) 235–258. https://doi.org/10.1177/0305829811419444cox.

Croft, S. (2006). *Culture, Crisis and America's War on Terror*. Cambridge University Press.

Czaputowicz, J. (2007). *Teorie stosunków międzynarodowych. Krytyka i systematyzacja*. Wydawnictwo Naukowe PWN. http://dx.doi.org/10.12775/HiP.2008.009.

Cześnik, M. (2011). Zawartość programów wyborczych polskich partii politycznych – zastosowanie metody MRG do ilościowej analizy treści. In I. Skłodkowska (Ed.). *Partie i ich programy polityczne. Wybory 2007* (pp. 9–25). Instytut Studiów Politycznych PAN.

Czyżewski, M., Kowalski, S., & Piotrowski, A. (1997). (Eds.). *Rytualny chaos. Studium dyskursu publicznego*. Wydawnictwo Aureus.

Derian, J. D. (2009). *The value of security: Hobbes, Marx, Nietzsche, and Baudrillard*. Routledge.

Deudney, D. (1990a). Environment and security: Muddled thinking. *Bulletin of the Atomic Scientists, 47*(3), 21–28. https://doi.org/10.1080/00963402.1991.11459957.

Deudney, D. (1990b). The case against linking environmental degradation and national security. Millennium:*Journal of International Studies, 19*(3), 461–476. https://doi.org/10.1177/03058298900190031001.

Dillon, M. (1996). *Politics of security: Towards a political philosophy of continental thought*. Routledge.

Diskaya, A. (2013, February 1). Towards a critical securitization theory: The Copenhagen and Aberystwyth schools of security studies. *E-International Relations*. http://www.e-ir.info/2013/02/01/towards-a-critical-securitizationtheory-the-copenhagen-and-aberystwyth-schools-ofsecurity-studies/#_ftn6.

Dittgen, H. & Peters, D. (2001, September 8–10). *EU and NATO: competing visions of security in Europe* [Conference session]. 4th Pan European International Relations Conference of the ECPR Standing Group on International Relations. University of Kent, Canterbury, United Kingdom.

Doty, R. L. (1993). Foreign Policy as Social Construction: A Post-Positivist Analysis of U.S. Counterinsurgency Policy in the Philippines. *International Studies Quarterly, 37*(3), 297–320. https://doi.org/10.2307/2600810.

Dourley, B. (2016, December 13). The "Riskification" of Climate Change: An Examination of Contemporary Practices of Climate Change Security. *ARST 629 Final Paper*. https://www.academia.edu/33381236/The_Riskification_of_Climate_Change_An_Examination_of_Contemporary_Practices_of_Climate_Change_Security.

Drabik, K. (Ed.). (2013). *Natura bezpieczeństwa w perspektywie personalnej i strukturalnej*. Akademia Obrony Narodowej.

Elbe, S. (2006). Should HIV/AIDS Be Securitized? The Ethical Dilemmas of Linking HIV/AIDS and Security. *International Studies Quarterly, 50*(1), 119–144. http://www.jstor.org/stable/3693554.

Ficoń, K. (2013). Bezpieczeństwo jako systemowa kategoria ontologiczna. *Bellona, 1*, 9–28. https://zbrojni.blob.core.windows.net/pzdata/TinyMceFiles/kwartalnik_bellona1_2013.pdf.

Fidler, D. P. & Gostin, L. (2008). *Biosecurity in the Global Age: Biological Weapons, Public Health and the Rule of Law*. Stanford University Press.

Fierke, K. M. (2007). *Critical Approaches to International Security*. Cambridge University Press.

Fijałkowski, Ł. (2012). Teoria sekurytyzacji i konstruowanie bezpieczeństwa. *Przegląd Strategiczny*, *1*, 149–161.

Fijałkowski, Ł. (2013). "Akt mowy" i "język bezpieczeństwa" a konstruowanie globalnych zagrożeń (dyskursywne podejście do bezpieczeństwa). Kultura – Historia – Globalizacja, *14*, 91–103.

Fijałkowski, Ł. & Jarząbek, J. (2016). Sekurytyzacja problemu uchodźców w Jordanii. In K. Zdulski, R. Czulda & R. Bania (Eds.). *Problemy bezpieczeństwa Bliskiego Wschodu i Północnej Afryki* (MENA) (pp. 30–47). Wydawnictwo Uniwersytetu Łódzkiego. http://dx.doi.org/10.18778/8088-164-8.03.

Fischer, D. (1993). *Nonmilitary Aspects of Security: A Systems Approach*. UNIDIR. United Nations.

Fischhendler, I. & Katz, D. (2013). The use of "security" jargon in sustainable development discourse: evidence from UN Commission on Sustainable Development. *International Environmental Agreements: Politics, Law and Economics*, *13*(3), 321–342. https://doi.org/10.1007/s10784-012-9192-z.

Floyd, R. (2007). Towards a consequentialist evaluation of security: bringing together the Copenhagen and the Welsh Schools of security studies. *Review of International Studies*, *33*(2), 327–350. https://doi.org/10.1017/S026021050700753X.

Floyd, R. (2008). *The Environmental Security Debate and Its Significance for Climate Change*. University of Warwick institutional repository. http://dx.doi.org/ 10.1080/03932720802280602.

Floyd, R. (2010). *Security and the environment: securitization theory and US environmental security policy*. Cambridge University Press.

Floyd, R. (2011). Can securitization theory be used in normative analysis? Towards a just securitization theory. *Security Dialogue*, *42*(4–5), 427–439. https://doi.org/10.1177/0967010611418712.

Floyd, R. (2012). Whither environmental security studies? An afterword. In R. Floyd & R. A. Matthew (Eds.). *Environmental Security: Approaches and issues* (pp. 279–297). Routledge.

Floyd, R. (2014). Just and unjust desecuritization. In T. Balzacq (Ed.). *Contesting Security: Strategies and Logics* (pp. 122–138). Routledge. https://doi.org/10.4324/9780203079850.

Floyd, R. (2016a). Extraordinary or ordinary emergency measures: what, and who, defines the 'success' of securitization? *Cambridge Review of International Affairs*, *29*(2), 677–694. https://doi.org/10.1080/09557571.2015.1077651.

Floyd, R. (2016b). The promise of theories of just securitization. In J. Nyman & A. Burke (Eds.). *Ethical Security Studies* (pp. 75–88). Routledge. https://doi-org.ep.fjernadgang.kb.dk/10.4324/9781315692005.

Floyd, R. (2019). *The Morality of Security*. Cambridge University Press.

Foucault, M. (2000). *Filozofia, historia, polityka. Wybór pism*. Wydawnictwo Naukowe PWN.

Foucault, M. (2005). *Słowa i rzeczy: Archeologia nauk humanistycznych*. Słowo/ obraz terytoria.

Foucault, M. (2006). Trzy typy władzy. In A. Jasińska-Kania, L. M. Nijakowski, J. Szacki & M. Ziółkowski (Eds.). *Współczesne teorie socjologiczne* (537–544). Wydawnictwo Naukowe Scholar.

Frankowski, P. (2005). Świat po zimnowojenny w świetle teorii chaosu. In R. Kuźniar (Ed.). *Porządek międzynarodowy u progu XXI wieku. Wizje – koncepcje – paradygmaty* (pp. 541–554). Wydawnictwa Uniwersytetu Warszawskiego.

Freedman, L. (1998). International security: Changing targets. *Foreign Policy, 110*, 48–63. https://doi.org/10.2307/1149276.

Frei, D. (1977). *Sicherheit: Grundfragen der Weltpolitik*. Verlag W. Kohlhammer.

Gad, U. P. & Petersen, K. L. (2011). Concepts of politics in securitization studies. *Security Dialogue, 42*(4–5), 315–328. https://doi.org/10.1177/0967010611418716.

Gajda, J. (2010). Analiza dyskursu i jej zastosowanie w badaniach politologicznych – wprowadzenie. In D. Mikucka-Wójtowicz (Ed.). *Przeszłość – Teraźniejszość – Przyszłość. Problemy badawcze młodych politologów* (pp. 253–261). Wydawnictwo LIBRON.

Gajda, S. (2005). Tekst/dyskurs oraz jego analiza i interpretacja. In M. Krauz & S. Gajda (Eds.). *Współczesne analizy dyskursu.* (pp. 11–20). Wydawnictwo Uniwersytetu Rzeszowskiego.

Garrett, L. (2005). *HIV and National Security: Where Are the Links?* Council on Foreign Relations. https://backend-live.cfr.org/sites/default/files/pdf/2005/07/HIV_National_Security.pdf.

Ginsberg, B. (1976). Elections and Public Policy, *American Political Science Review, 70*(1), 41–49. https://doi.org/10.1017/S0003055400263971.

Glen, A. (2011). Podstawy poznawcze bezpieczeństwa narodowego. *Zeszyty Naukowe. Akademia Obrony Narodowej, 2*(83), 19–25.

Graeger, N. (1996). Environmental security? *Journal of Peace Research, 33*(1), 109–116. https://doi.org/10.1177/0022343396033001008.

Grinth, H. (1996). Texte im politischen diskurs. Ein Vorschlag zu diskursorientieren Beschreibung von Textsorten, *Muttersprache, 106*, 66–80.

Grzegorczykowa, R. (2008). *Wstęp do językoznawstwa*. Wydawnictwo Naukowe PWN.

Grzymała-Kazłowska, A. (2014). Socjologicznie zorientowana analiza dyskursu na tle współczesnych badań nad dyskursem. *Kultura i Społeczeństwo, 48*(1), 13–34.

Habermas, J. (1986). Pojęcie działania komunikacyjnego: uwagi wyjaśniające. *Kultura i Społeczeństwo, 30*(3), 21–44.

Habermas, J. (2002). *Teoria działania komunikacyjnego.* Wydawnictwo Naukowe PWN.

Haftendorn, H., Keohane, R. O., & Wallander, C. A. (1999). *Imperfect unions: security institutions over time and space.* Oxford University Press.

Hale, S. (2010). The new politics of climate change: Why are we falling and how we can succeed. *Environmental Politics, 19*(2), 255–275. https://doi.org/10.1080/096440 10903576900.

Hanlon, R. J. & Christie, K. (2016). *Freedom from Fear, Freedom from Want: An Introduction to Human Security.* University of Toronto Press.

Hansen, L. (2000). The Little Mermaid's silent security dilemma and the absence of gender in the Copenhagen School. *Millennium: Journal of International Studies, 29*(2), 285–306. https://doi.org/10.1177/03058298000290020501.

Hansen, L. (2012). Reconstructing desecuritisation: the normative-political in the Copenhagen School and directions for how to apply it. *Review of International Studies, 38*(3), 525–546. doi:10.1017/S0260210511000581.

Hardy, C., & MaGuire, S. (2016). Organizing risk: Discourse, power, and "riskification." *Academy of Management Review, 41*(1), 80–108. https://doi.org/10.5465/amr .2013.0106.

Hastoro, D. & Sumarni, D. W. (2020). Human security, social stigma, and global health: the COVID-19 pandemic in Indonesia. *Journal of the Medical Sciences, 52*(3)(SI), 158–165. http://dx.doi.org/10.19106/JMedSciSI005203202014.

Haughton, S. A. (2023). Securitization and De-securitization. In: S. N. Romaniuk & P. Marton (Eds.), *The Palgrave Encyclopedia of Global Security Studies* (p. 1297–1299). Palgrave Macmillan. https://doi.org/10.1007/978-3-319-74319-6.

Heck, A. & Schlag, G. (2013). Securitizing images: The female body and the war in Afghanistan. *European Journal of International Relations 19*(4), 891–913. https://doi .org/10.1177/1354066111433896.

Herington, J. (2012). The Concept of Security. In M. Selgelid & Ch. Enemark (Eds.), *Ethical and Security Aspects of Infectious Disease Control: Interdisciplinary Perspectives.* (pp. 7–25). Ashgate Publishing.

Hołyst, B. (2014). *Bezpieczeństwo. Ogólne problemy badawcze.* Wydawnictwo Naukowe PWN.

Hoogensen, G. & Rottem, S. V. (2004). Gender identity and the subject of security. *Security Dialogue, 35*(2), 155–171. https://doi.org/10.1177/0967010604044974.

Howell, A., & Richter-Montpetit, M. (2020). Is securitization theory racist? Civilizationism, methodological whiteness, and antiblack thought in the Copenhagen School. *Security Dialogue, 51*(1), 3–22. https://doi.org/10.1177/0967010619862921.

Hudson, H. (2005). 'Doing' security as though humans matter: A feminist perspective on gender and the politics of human security. *Security Dialogue, 36*(2), 155–174. https://doi.org/10.1177/0967010605054642.

Hudson, N. F. (2009). Securitizing women's rights and gender equality. *Journal of Human Rights, 8*(1), 53–70. https://doi.org/10.1080/14754830802686526.

Huysmans, J. (1995). Migrants as a Security Problem: Dangers of 'Securitizing' Societal Issues. In R. Miles & D. Thränhardt (Eds.), *Migration and European integration: the dynamics of inclusion and exclusion* (pp. 53–72). Fairleigh Dickinson University Press, Pinter Publishers.

Huysmans, J. (1998). The Question of the Limit: Desecuritisation and the Aesthetics of Horror in Political Realism. *Millennium: Journal of International Studies, 27*(3), 569–589. https://doi.org/10.1177/03058298980270031301.

Huysmans, J. (2000). Migration and the Politics of Security. In S. Body-Gendrot & M. Martiniello (Eds.), *Minorities in European Cities. Migration, Minorities and Citizenship* (pp. 179–189). Palgrave Macmillan. https://doi.org/10.1007/978-1-349-62841-4_13.

Huysmans, J. (2006). *The Politics of Insecurity: Fear, Migration and Asylum in the EU.* Routledge.

Huysmans, J. (2014). *Security Unbound: Enacting Democratic Limits.* Routledge.

Ibrahim, M. (2005). The securitization of migration: A racial discourse. *International Migration, 43*(5), 163–187. https://doi.org/10.1111/j.1468-2435.2005.00345.x.

Jakubczak, R. & Flis, J. (Eds.). (2006). *Bezpieczeństwo narodowe Polski w XXI wieku. Wyzwania i strategie.* Dom Wydawniczy Bellona.

Judge, A. & Maltby, T. (2017). European Energy Union? Caught between securitisation and 'riskification.' *European Journal of International Security, 2*(2), 179–202. https://doi.org/10.1017/eis.2017.3.

Jutila, M. (2006). Desecuritizing Minority Rights: Against Determinism. *Security Dialogue, 37*(2), 167–186. https://doi.org/10.1177/0967010606066169.

Kaczmarek, J. (2008). *Współczesne bezpieczeństwo.* Wydawnictwo AON.

Kearns, M. (2017). Gender, visuality and violence: Visual securitization and the 2001 war in Afghanistan. *International Feminist Journal of Politics, 19*(4), 1–15. https://doi.org/10.1080/14616742.2017.1371623.

Kitler, W. (2010). Bezpieczeństwo narodowe. Podstawowe kategorie, dylematy pojęciowe i próba systematyzacji. *Zeszyt Problemowy TWO, 1*(61).

Klotz, L. & Sylvester, E. J. (2009). *Breeding Bio Insecurity: How US Biodefense is Exporting Fear, Globalizing Risk, and Making Us All Less Secure.* University of Chicago Press.

Koblenz, G. D. (2010). Biosecurity reconsidered: Calibrating biological threats and responses. *International Security*, *34*(4), 96–132. https://doi.org/10.1162/isec.2010.34.4.96.

Kolodziej, E. A. (1992). Renaissance in Security Studies? Caveat Lector! *International Studies Quarterly*, *36*(4), 421–438. https://doi.org/10.2307/2600733.

Kopaliński, W. (1983). *Słownik języka polskiego*. Wiedza Powszechna.

Kostecki, W. (2012). *Strach i potęga. Bezpieczeństwo międzynarodowe w XXI wieku*. Wydawnictwo Poltext.

Krause, K. & Williams, M. C. (1996). Broadening the Agenda of Security Studies: Politics and Methods. *Mershon International Studies Review*, *40*(2), 229–254. https://doi.org/10.2307/222776.

Krause, K. & Williams, M. C. (1997). *Critical Security Studies: Concepts and Cases* (*1st ed.*). Routledge. https://doi-org.ep.fjernadgang.kb.dk/10.4324/9780203501764.

Krause, K. & Williams, M. C. (2018). Security and "Security Studies": Conceptual Evolution and Historical Transformation. In A. Gheciu & W. C. Wohlforth (Eds.), *The Oxford Handbook of International Security* (pp. 14–28). Oxford Academic. https://doi.org/10.1093/oxfordhb/9780198777854.013.2.

Krippendorff, K. (2004). Measuring the Reliability of Qualitative Text Analysis Data. *Quality and Quantity*, *38*(6), 787–800. https://doi.org/10.1007/s11135-004-8107-7.

Kruyt, B., van Vuuren, D. P., de Vries, H. J. M., & Groenenberg, H. (2009). Indicators for energy security. *Energy Policy*, *37*(6), 2166–2181. http://dx.doi.org/10.1016/j. enpol.2009.02.006.

Kujawa, I. (2009). Strategie dyskursu polityki a medialny obraz świata (na materiale współczesnej prasy niemieckiej), *Annales Universitatis Mariae Curie-Sklodowska*, *27*(FF), 43–58.

Kukułka, J. (1982). Bezpieczeństwo a współpraca europejska: współzależności i sprzeczności interesów. *Sprawy międzynarodowe*, *7*, 45–57.

Kukułka, J. (Ed.). (1994). *Bezpieczeństwo międzynarodowe w Europie Środkowej po zimnej wojnie*. Wydawnictwo Naukowe Scholar.

Lake, D. (2001). Beyond anarchy: the importance of security institutions, *International Security*, *26*(1), 129–160.

Leech, G. N. & Short, M. H. (1981). *Style in Fiction: A Linguistic Introduction to English Fictional Prose*. Longman.

Leonard, S. & Kaunert, C. (2011). Reconceptualizing the audience in securitization theory. In T. Balzacq (ed.), *Securitization theory: How security problems emerge and dissolve* (pp. 57–76). Routledge.

Levy, M. (1995). Is the environment a national security issue? *International Security*, *20*(2), 35–62. https://doi.org/10.2307/2539228.

Lisiecki, M. & Kwiatkowska-Basałaj, B. (2000). Pojęcie bezpieczeństwa oraz prognostyczny model jego zapewnienia. In P. Tyrała (Ed.), *Zarządzanie bezpieczeństwem* (pp. 53–60). Wydawnictwo Profesjonalnej Szkoły Biznesu.

Manifesto Research on Political Representation, MARPOR 2020. https://manifesto-project.wzb.eu.

Markowski, R. (2002). Propozycja *Manifesto Research Group*: metoda, wyniki, problemy – komentarz. In R. Markowski (Ed.), *System partyjny i zachowania wyborcze: dekada polskich doświadczeń* (289–297). ISP PAN and Ebert Stiftung.

McDonald, M. (2008). Securitization and the construction of security. *European Journal of International Relations*, *14*(4), 563–587. https://doi.org/10.1177/1354066108097553.

McDonald, M. (2012). Konstruktywizm, In P. D. Williams (Ed.), *Studia bezpieczeństwa* (pp. 59–72). Wydawnictwo Uniwersytetu Jagiellońskiego.

McInnes, C. & Rushton, S. (2013). HIV/AIDS and securitization theory. *European Journal of International Relations*, *19*(1), 115–138. https://doi.org/10.1177/1354066111425258.

Meyer, M. (2001). Between theory, method and politics: positioning of the approaches to CDA. In M. Meyer & R. Wodak (Eds.), *Methods of Critical Discourse Analysis* (14–32). Sage Publications.

Michael, K. (2009). Who really dictates what an existential threat is? The Israeli experience. *Journal of Strategic Studies*, *32*(5), 687–713.

Mikiewicz, P. (2005). *Kategoria bezpieczeństwa a polska myśl polityczna lat 90*. Arboretum.

Mofette, D. & Vadasaria, S. (2016). Uninhibited violence: Race and the securitization of immigration. *Critical Studies on Security*, *4*(3), 291–305. https://doi.org/10.1080/21624887.2016.1256365.

Morgenthau, H. J., Thompson, K. W., & Clinton, W. D. (2006). *Politics among nations: The struggle for power and peace*. Boston: McGraw-Hill Higher Education.

Musioł, M. (2010). Znaczenie sekurytyzacji i sektorów bezpieczeństwa w ramach krytycznych studiów nad bezpieczeństwem. *Historia i Polityka*, *23*(30), 39–51. http://dx.doi.org/10.12775/HiP.2018.003.

Navari, C. (2012). Liberalizm. In P. D. Williams (Ed.), *Studia bezpieczeństwa* (pp. 29–43). Wydawnictwo Uniwersytetu Jagiellońskiego.

Neack, L. (2017). *National, International, and Human Security: A Comparative Introduction*. Rowman & Littlefield.

Nurhasanah, S., Napang, M. & Rohman, S. (2020). Covid-19 As A Non-Traditional Threat To Human Security. *Journal of Strategic and Global Studies*, *3*(1). Article 5. https://doi.org/10.7454/jsgs.v3i1.1028.

Oels, A. (2012). From 'Securitization' of Climate Change to 'Climatization' of the Security Field: Comparing Three Theoretical Perspectives. In J. Scheffran, M. Brzoska,

H. Brauch, P. Link & J. Schilling (Eds.), *Climate Change, Human Security and Violent Conflict, Hexagon Series of Human and Environmental Security and Peace*, 8 (pp. 185–205). Springer. https://doi.org/10.1007/978-3-642-28626-1_9.

Pietraś, M. (2007). (Ed.). *Międzynarodowe stosunki polityczne*. Wydawnictwo UMCS.

Plumpe, G. (1988). Kunst und juristischer Diskurs. Mit einer Vorbemerkung zum Diskursbegriff. In J. Fohrmann & H. Müller (Eds.), *Diskurstheorien und Literaturwissenschaft* (pp. 330–345). Suhrkamp.

Prins, G. (1998). *Environmental change and international security*. Rijksuniversiteit Groningen, Vakgroep Bestuursrecht en Bestuurskunde.

Pytkowski, W. (1985). *Organizacja badań i ocena prac naukowych*. Wydawnictwo Naukowe PWN.

Ricoeur, P. (1989). *Język, tekst, interpretacja*. Państwowy Instytut Wydawniczy.

Rittel, S. J. (2005). *Dyskurs w filozofii politycznej. Podejście lingwistyczno-politologiczne i systemowe*. Wydawnictwo Akademii Świętokrzyskiej im. Jana Kochanowskiego.

Robertson, D. A. (1976). *Theory of Party Competition*. Wiley.

Robinson, D. (2010). Critical security studies and the deconstruction of realist hegemony. *Journal of Alternative Perspectives in the Social Sciences*, *2*(2), 846–853.

Roe, P. (2004). Securitization and minority rights: Conditions of desecuritization. *Security Dialogue*, *35*(3), 279–294. https://doi.org/10.1177/0967010604047527.

Roe, P. (2008). Actor, Audience(s) and Emergency Measures: Securitization and the UK's Decision to Invade Iraq. *Security Dialogue*, *39*(6), 615–635. https://doi.org/10.1177/0967010608098212.

Rothschild, E. (1995). *The debate on economic and social security in the late eighteenth century: Lessons of a road not taken*. United Nations Research Institute for Social Development.

Rummel, R. J. (1994). Power, Genocide and Mass Murder. *Journal of Peace Research*, *31*(1), 1–10. https://doi.org/10.1177/0022343394031001001.

Russell, B. (1997). *Władza i jednostka*. Książka i Wiedza.

Sabaratnam, M. (2013). Avatars of Eurocentrism in the critique of the liberal peace. *Security Dialogue*, *44*(3), 259–278. https://doi.org/10.1177/0967010613485870.

Salter, M. B. (2008). Securitization and de-securitization: A dramaturgical analysis of the Canadian Air Transport Security Authority. *Journal of International Relations and Development*, *11*, 321–349.

Schmitt, C. (1985). *Political theology: Four chapters on the concept of sovereignty*. MIT Press.

Ścibiorek, Z., Wiśniewski, B. Kuc, R. B. & Dawidczyk, A. (2017). *Bezpieczeństwo wewnętrzne. Podręcznik akademicki*. Wydawnictwo Adam Marszałek.

Seth, J., Bakker, R., Hooghe, L., Marks, G., Polk, J., Rovny, J., Steenbergen, M., & Vachudova, M. A. (2022). Chapel Hill Expert Survey Trend File, 1999–2019. *Electoral Studies, 75* (February). https://doi.org/10.1016/j.electstud.2021.102420.

Shani, G. (2020). Securitizing 'Bare Life'? Human Security and Coronavirus. *E-International Relations.* https://www.e-ir.info/2020/04/03/securitizing-bare-life-human-security-and-coronavirus/.

Shiping, T. (2010). Offence–defence Theory: Towards a Definitive Understanding. *The Chinese Journal of International Politics, 3*, 213–260.

Smith, S. (2005). The Contested Concept of Security. In K. Booth (Ed.), *Security Studies and World Politics.* Lynne Rienner Publishers.

Stańczyk, J. (2009). Złożoność kategorii bezpieczeństwa – zarys problematyki. In E. Jarmocha, A. W. Świderski & I. A. Trzpil (Eds.). *Bezpieczeństwo człowieka a transdyscyplinowość.* Tom II *Aspekty edukacyjne i politologiczne* (pp. 239–255). Wydawnictwo Akademii Podlaskiej.

Stańczyk, J. (2011). Istota współczesnego pojmowania bezpieczeństwa – zasadnicze tendencje. *Rocznik Bezpieczeństwa Międzynarodowego, 2010/2011*, 15–33.

Stańczyk, J. (2017). *Formułowanie kategorii pojęciowej bezpieczeństwa.* FNCE sp. z o.o.

Stefanowicz, J. (1984). *Bezpieczeństwo współczesnych państw.* Pax.

Stern, M., & Ojendal, J. (2010). Mapping the security – Development nexus: Conflict, complexity, cacophony, convergence? *Security Dialogue, 41*(1), 5–29. https://doi.org/10.1177/0967010609357041.

Strauss, G., Hass, U. & Harras, G. (1989). *Brisante Wörter von Agitation bis Zeitgeist.* De Gruyter.

Stritzel, H. (2007). Towards a theory of securitization: Copenhagen and Beyond. *European Journal of International Relations, 13*(3), 357–383. https://doi.org/10.1177/1354066107080128.

Strzelczyk, J. (2002). *Ucieczka ze Wschodu. Rosja w polskiej polityce 1989–1993.* Rytm.

Świniarski, J. (1997). *O naturze bezpieczeństwa: prolegomena do zagadnień ogólnych.* Agencja Wydawnicza ULMAK.

Świniarski, J. (1999). *Filozoficzne podstawy edukacji dla bezpieczeństwa.* Departament Społeczno-Wychowawczy MON. Agencja Wydawnicza "Egros."

Synder, G. H. (2007). *Alliance Politics.* Ithaca, Cornell University Press.

Szacki, J. (2002). *Historia myśli socjologicznej.* Wydawnictwo Naukowe PWN.

Takao, T. (2020). Japan's Leadership in Human Security During and After the COVID-19 Pandemic, *Asia-Pacific Review, 27*(2), 26–45. https://doi.org/10.1080/13439006.2020.1841950.

Tarlock, D. & Wouters, P. (2009). Reframing the water security dialogue. *Water Law, 20*, 53–60.

Taureck, R. (2006). Securitization theory and securitization studies. *Journal of International Relations and Development*, *9*(1), 53–61.

Trombetta, M. J. (2008). Environmental security and climate change: Analyzing the discourse. *Cambridge Review of International Affairs*, *21*(4), 585–602. https://doi.org/10.1080/09557570802452920.

Trombetta, M. J. (2011). Rethinking the securitization of the environment: Old beliefs, new insights. In T. Balzacq (Ed.), *Securitization theory: How security problems emerge and dissolve* (pp. 135–149). Routledge.

Trutkowski, C. (2004). Wybór czy konieczność – o potrzebie wykorzystania analizy dyskursu w socjologii. *Kultura i Społeczeństwo*, *48*(1), 35–50.

Urbanek, A. (2013). (Ed.) *Wybrane problemy bezpieczeństwa*. Wydawnictwo Społeczno-Prawne.

van Dijk, T. A. (2003). *Political discourse and ideology*. Retrieved from: http://hdl.handle.net/10637/6038.

van Munster, R. (2005). Logics of Security: The Copenhagen School, Risk Management and the War on Terror. *Political Science Publications*, *10*(6). https://www.sdu.dk/~/media/Files/Om_SDU/Institutter/Statskundskab/Skriftserie/05RVM10.as.

Vedby Rasmussen, M. (2004). 'It Sounds Like a Riddle': Security Studies, the War on Terror and Risk. *Millennium*, *33*(2), 381–395. https://doi-org.ep.fjernadgang.kb.dk/10.1177/03058298040330020601 (Original work published 2004).

Vedby Rasmussen, M. (2006). *The Risk Society at War. Terror, Technology and Strategy in Twenty-First Century*. Cambridge University Press.

Villarreal, P. (2021, February 26). The Security Council and COVID-19 – Towards a Medicalization of International Peace and Security. *ESIL Reflections*, *9*(6). European Society of International Law. https://esil-sedi.eu/wp-content/uploads/2021/02/ESIL-Reflection-Villarreal-1-2.pdf.

Vogler, J. (2013). International Relations Theory and the Environment. In G. Ktting (Ed.), *Global Environmental Politics, Concepts, Theories and Case Studies* (pp. 11–26). Routledge.

Volkens, A., Bara, J., & Budge, I. (2009). Data Quality in Content Analysis. The Case of the Comparative Manifestos Project. *Historical Social Research*, *34*, (1/127), 234–251. http://www.jstor.org/stable/20762343.

von Lucke, F. (May 2015). The Attempted Securitisation of Climate Change in Mexico: Explorations into Actors, Processes and Consequences. *ClimaSec Working Paper Series*, *9*. University of Tübingen.

von Lucke, F., Wellmann, Z. & Diez, T. (2014). What's at Stake in Securitising Climate Change? Towards a Differentiated Approach. *Geopolitics*, *19*(4), 857–884. https://doi.org/10.1080/14650045.2014.913028.

Vuori, J. (2008). Illocutionary logic and strands of securitization: Applying the theory of securitization to the study of non-democratic political orders. *European Journal of International Relations, 14*(1), 65–99. https://doi.org/10.1177/1354066107087767.

Walt, S. M. (1991). The Renaissance of Security Studies. *International Studies Quarterly, 35*(2), 211–239. https://doi.org/10.2307/2600471.

Waltz, K. N. (1959). *Man, the State and War. A theoretical analysis*. Columbia University Press.

Wæver, O. (1993). Societal Security: the concept. In O. Wæver, B. Buzan, M. Kelstrup & P. Lemaitre (Eds.), *Identity, Migration and New Security Agenda in Europe* (pp. 17–40). Pinter.

Wæver, O. (1995). Securitization and desecuritization. In R. D. Lipschutz (Ed.), *On Security* (pp. 46–86). Columbia University Press.

Wæver, O. (1997). *Concepts of Security*. Copenhagen Institute for Political Science, University of Copenhagen.

Wæver, O. (1998). Insecurity, security and asecurity in the Western European non-war community. In E. Adler & M. Barnett (Eds.), *Security Communities* (pp. 69–118). Cambridge: Cambridge University Press. https://doi.org/10.1017/CBO9780511598661.003.

Wæver, O. (2001). The EU as a security actor. Reflections from a pessimistic constructivist on post-sovereign security orders. In: M. Kelstrup & M. Williams (eds.). International Relations Theory and the Politics of European Integration. Power, Security and Community. (p. 250–294). Routledge.

Wæver, O. (2004, March 17). *Aberystwyth, Paris, Copenhagen. New Schools in Security Theory and their Origins between Core and Periphery* [Conference session]. International Studies Association, Montreal, Canada.

Wæver, O., Buzan, B., Kelstrup, M. & Lamaitre, P. (1993). *Identity, Migration and the New Security Agenda in Europe*. Pinter.

Weldes, J., Laffey, M., Gusterson, H. & Duvall, R. (1999). Introduction. In J. Weldes, M. Laffey, H. Gusterson & R. Duvall (Eds.), *Cultures of Insecurity: States, Communities, and the Production of Danger* (pp. 1–33). University of Minnesota Press.

Wendell, C. W. & Romaniuk, S. N. (2023). Critical Security Studies. In S. N. Romaniuk &P. Marton (Eds.), *The Palgrave Encyclopedia of Global Security Studies* (pp. 256–260). Palgrave Macmillan. https://doi.org/10.1007/978-3-319-74319-6.

Wendt, A. (2014). *Social theory of international politics*. New York: Cambridge University Press.

Wertman, O. & Kaunert, C. (2022). The Audience in Securitization Theory. *Strategic Assessment, 25*(3), 67–81.

Wilkinson, C. (2007). The Copenhagen School on tour in Kyrgyzstan: Is securitization theory useable outside Europe? *Security Dialogue, 38*(1), 5–25. https://doi.org/10.1177/0967010607075964.

Williams, M. C. (2003). Words, images, enemies: Securitization and international politics. *International Studies Quarterly*, *47*(4), 511–531.

Williams, P. D. (Ed.). (2012). *Studia bezpieczeństwa*. Wydawnictwo Uniwersytetu Jagiellońskiego.

Williams, R. E., Jr. & Viotti, P. R. (2012). *Arms Control: History, Theory, and Policy*. ABC-CLIO.

Wodak, R. (2001). The discourse-historical approach. In M. Meyer & R. Wodak (Eds.), *Methods of Critical Discourse Analysis* (pp. 87–121). Sage Publications.

Wodak, R. (2008). Dyskurs populistyczny: retoryka wykluczenia a gatunki języka pisanego, In A. Duszak & N. Fairclough (Eds.), *Krytyczna analiza dyskursu. Interdyscyplinarne podejście do komunikacji społecznej* (pp. 185–215), Towarzystwo Autorów i Wydawców Prac Naukowych UNIVERSITAS.

Wodak, R. (2011). Badania nad dyskursem – ważne pojęcia i terminy. In R. Wodak, M. Krzyżanowski (Eds.), *Jakościowa analiza dyskursu w naukach społecznych*. Oficyna Wydawnicza Łośgraf.

Wodak, R. & Krzyżanowski, M. (2017). (Eds.). Qualitative Discourse Analysis in the Social Sciences. Bloomsbury Academic. http://dx.doi.org/10.1007/978-1-137-04798-4&gt.

Wojtaszczyk, K. A. (2002). (Ed.). *Bezpieczeństwo Polski w perspektywie członkostwa w Unii Europejskiej*. Elipsa Dom Wydawniczy i Handlowy.

Wolanin, J. (2005). *Zarys teorii bezpieczeństwa obywateli. Ochrona ludności podczas pokoju*. DANMAR.

Wolfers, A. (1952). "National Security" as an Ambiguous Symbol. *Political Science Quarterly*, *67*(4), 481–502. https://doi.org/10.2307/2145138.

Woźniak, W. (2017). Programy polskich partii politycznych jako materiał empiryczny w badaniach dyskursu politycznego, *Środkowoeuropejskie Studia Polityczne*, *2*, 41–58. https://doi.org/10.14746/ssp.2017.2.3.

Zięba, R. (1999). *Instytucjonalizacja bezpieczeństwa europejskiego: koncepcje – struktury – funkcjonowanie*. Fundacja Studiów Międzynarodowych. Wydawnictwo Naukowe Scholar.

Zięba, R. (2006). Teoria ogólna bezpieczeństwa państwa w stosunkach międzynarodowych. In E. Haliżak, R. Kuźniar, G. Michałowska, S. Parzymies, J. Symonides & R. Zięba (Eds.), *Stosunki międzynarodowe w XXI wieku. Księga jubileuszowa z okazji 30-lecia Instytutu Stosunków Międzynarodowych Uniwersytetu Warszawskiego* (pp. 985–1004). Wydawnictwo Naukowe Scholar.

Zięba, R. (red.). (2008). *Bezpieczeństwo międzynarodowe po zimnej wojnie*. Wydawnictwa Akademickie i Profesjonalne.

Zięba, R. (2012). O tożsamości nauk o bezpieczeństwie. *Zeszyty Naukowe AON*, *1*(86), 7–22.

Index

www.ingramcontent.com/pod-product-compliance
Lightning Source LLC
Chambersburg PA
CBHW071247230926
51849CB00001B/252
* 9 7 8 9 0 0 4 7 4 5 4 0 7 *